MATHEMATICS RESEARCH DEVELOPMENTS

ASYMPTOTIC BEHAVIOR

AN OVERVIEW

MATHEMATICS RESEARCH DEVELOPMENTS

Additional books and e-books in this series can be found on Nova's website under the Series tab.

MATHEMATICS RESEARCH DEVELOPMENTS

ASYMPTOTIC BEHAVIOR

AN OVERVIEW

STEVE P. RILEY
EDITOR

Copyright © 2020 by Nova Science Publishers, Inc.

All rights reserved. No part of this book may be reproduced, stored in a retrieval system or transmitted in any form or by any means: electronic, electrostatic, magnetic, tape, mechanical photocopying, recording or otherwise without the written permission of the Publisher.

We have partnered with Copyright Clearance Center to make it easy for you to obtain permissions to reuse content from this publication. Simply navigate to this publication's page on Nova's website and locate the "Get Permission" button below the title description. This button is linked directly to the title's permission page on copyright.com. Alternatively, you can visit copyright.com and search by title, ISBN, or ISSN.

For further questions about using the service on copyright.com, please contact:
Copyright Clearance Center
Phone: +1-(978) 750-8400   Fax: +1-(978) 750-4470   E-mail: info@copyright.com.

NOTICE TO THE READER

The Publisher has taken reasonable care in the preparation of this book, but makes no expressed or implied warranty of any kind and assumes no responsibility for any errors or omissions. No liability is assumed for incidental or consequential damages in connection with or arising out of information contained in this book. The Publisher shall not be liable for any special, consequential, or exemplary damages resulting, in whole or in part, from the readers' use of, or reliance upon, this material. Any parts of this book based on government reports are so indicated and copyright is claimed for those parts to the extent applicable to compilations of such works.

Independent verification should be sought for any data, advice or recommendations contained in this book. In addition, no responsibility is assumed by the Publisher for any injury and/or damage to persons or property arising from any methods, products, instructions, ideas or otherwise contained in this publication.

This publication is designed to provide accurate and authoritative information with regard to the subject matter covered herein. It is sold with the clear understanding that the Publisher is not engaged in rendering legal or any other professional services. If legal or any other expert assistance is required, the services of a competent person should be sought. FROM A DECLARATION OF PARTICIPANTS JOINTLY ADOPTED BY A COMMITTEE OF THE AMERICAN BAR ASSOCIATION AND A COMMITTEE OF PUBLISHERS.

Additional color graphics may be available in the e-book version of this book.

Library of Congress Cataloging-in-Publication Data

Names: Riley, Steve P., editor.
Title: Asymptotic Behavior: An Overview
Description: New York: Nova Science Publishers, [2019] | Series: Mathematics Research Developments | Includes bibliographical references and index.
Identifiers: LCCN 2019957631 (print) | ISBN 9781536172225 (paperback) |
  ISBN 9781536172232 (adobe pdf)

Published by Nova Science Publishers, Inc. † New York

CONTENTS

Preface		**vii**
Chapter 1	Oscillation Criteria for First Order Partial Dynamic Equations with Several Delays on Time Scales *Svetlin G. Georgiev*	**1**
Chapter 2	Asymptotic Behavior in Quantum-Field Models from Schwinger-Dyson Equations *V. E. Rochev*	**53**
Chapter 3	Asymptotic Behavior for the Hydrogen Atom Confined by Different Potentials *Michael-Adán Martínez-Sánchez, Rubicelia Vargas and Jorge Garza*	**101**
Index		**133**
Related Nova Publications		**137**

PREFACE

Asymptotic Behavior: An Overview is designed to provide the reader with an exposition of some aspects of the oscillation theory of first order delay partial dynamic equations on time scales. Oscillation theory of differential equations, originated from the monumental paper of C. Sturm published in 1836, has now been recognized as an important branch of mathematical analysis from both theoretical and practical viewpoints.

Asymptotic behavior in the deep Euclidean region of momenta for four-dimensional models of quantum field theory is studied through the system of Schwinger-Dyson equations. This system is truncated by a sequence of n-particle approximations in which $n \to \infty$ goes into the complete system of Schwinger-Dyson equations.

Lastly, the authors discuss the exact analytical solution of the Schrdinger equation corresponding to the hydrogen atom confined by four spherical potentials: infinite potential, parabolic potential, constant potential, and dielectric continuum.

As shown in Chapter 1, oscillation theory of differential equations, originated from the monumental paper of C. Sturm published in 1836, has now been recognized as an important branch of mathematical analysis from both theoretical and practical viewpoints. It has been attracting wide attention of researchers for the past five decades, and as a consequence of their great efforts it has grown up to be a fertile field covering ordinary differential equations, functional differential equations with deviating arguments, partial differential equations with or without functional arguments and difference equations. It goes without saying that a vast literature, research papers and books, dealing with oscillation theory has been published so far. As regards books and monographs on the subject, most of the existing ones are exclusively concerned with ordinary differential equations and/or difference equations, and there seems to be none which is devoted to the study of oscillations of partial differential equations. Chapter 1 is designed to provide the reader with an exposition of some aspects of the oscillation theory of first order delay partial dynamic equations on time scales. It is investigated a class BVP of first order delay partial dynamic equations on time scales for existence of solutions and then they are deducted some oscillations criteria for first order partial dynamic equations on time scales.

In Chapter 2, asymptotic behavior in the deep Euclidean region of momenta for four-dimensional models of quantum field theory is studied by using the system of Schwinger-Dyson equations (SDEs). This system is truncated by a sequence of n-particle approximations which for $n \to \infty$ goes into the complete system of SDEs. For a model of complex scalar field ϕ with interaction $\frac{\lambda}{2}(\phi^*\phi)^2$ an asymptotic solution of the system of SDEs in two-particle approximation is obtained. The two-particle amplitude has the pathology-free asymptotic behavior at large momenta in the region of strong coupling at $\lambda > \lambda_{cr}$. At $\lambda < \lambda_{cr}$ this amplitude possesses Landau-type singularity. For pseudoscalar Yukawa model a solution of the two-particle approximation for the pseudoscalar propagator is free from non-physical singularities and has the self-consistent asymptotic behavior. The investigation of the super-renormalized model of a complex scalar field ϕ and a real scalar field χ with the interaction $g\phi^*\phi\chi$ demonstrates a change of the asymptotic behavior in the Euclidean region of momenta in a vicinity of a certain critical value of the coupling constant. For small values of the coupling the propagator of field ϕ behaves asymptotically as free. In the strong-coupling region the asymptotic behavior drastically changes – the propagator in the deep Euclidean region tends to a constant. In the coordinate space this propagator has a characteristic shell structure. The same shell structure in coordinate space has a vertex with zero momentum transfer. An analogy between the phase transition in this model and the re-arrangement of the physical vacuum in the supercritical external field due to the "fall-on-the-center" phenomenon is discussed.

The exact wave function of the *free* hydrogen atom has been used to design basis sets employed for the study of many-electron atoms; however, when we take into account the impact of the environment in which is immersed an atom, the behavior of the wave function is different from that showed by the free atom. In Chapter 3, the authors discuss the exact analytical solution of the Schrödinger equation corresponding to the hydrogen atom confined by four spherical potentials: I) Infinite potential. II) Parabolic potential. III) Constant potential. IV) Dielectric continuum. All these potentials are applied in the region $r \geq r_0$ where r_0 constitutes a confinement radius. In all cases, the potential $-Z/r$ is defined in the region $r < r_0$. As a general conclusion we found that the wave function decays faster in the following ordering: Pot. I > Pot. II > Pot. III > Pot. IV. The analytical expression for the exact wave function shows how the asymptotic behavior is for each potential and consequently this information suggests how to build a basis set to study the electronic structure of many-electron atoms under the same spatial restrictions. Thus, the authors present results when Gaussian functions are used as a basis set in the Ritz method and we contrast its results with those obtained by the exact solution for the confined hydrogen atom.

In: Asymptotic Behavior: An Overview
Editor: Steve P. Riley

ISBN: 978-1-53617-222-5
© 2020 Nova Science Publishers, Inc.

Chapter 1

OSCILLATION CRITERIA FOR FIRST ORDER PARTIAL DYNAMIC EQUATIONS WITH SEVERAL DELAYS ON TIME SCALES

Svetlin G. Georgiev[*]
Sorbonne University, Paris, France

Abstract

Oscillation theory of differential equations, originated from the monumental paper of C. Sturm published in 1836, has now been recognized as an important branch of mathematical analysis from both theoretical and practical viewpoints. It has been attracting wide attention of researchers for the past five decades, and as a consequence of their great efforts it has grown up to be a fertile field covering ordinary differential equations, functional differential equations with deviating arguments, partial differential equations with or without functional arguments and difference equations. It goes without saying that a vast literature, research papers and books, dealing with oscillation theory has been published so far. As regards books and monographs on the subject, most of the existing ones are exclusively concerned with ordinary differential equations and/or difference equations, and there seems to be none which is devoted to the study of oscillations of partial differential equations. The present chapter is designed to provide the reader with an exposition of some aspects of the oscillation theory of first order delay partial dynamic equations on time

---
[*]Corresponding Author's E-mail: svetlingeorgiev1@gmail.com.

scales. It is investigated a class BVP of first order delay partial dynamic equations on time scales for existence of solutions and then they are deducted some oscillations criteria for first order partial dynamic equations on time scales.

1. INTRODUCTION

Oscillation theory of differential equations, originated from the monumental paper of C. Sturm published in 1836, has now been recognized as an important branch of mathematical analysis from both theoretical and practical viewpoints.

Roughly speaking, the objective of oscillation theory is to acquire as much information as possible about the qualitative properties of solutions of differential equations through the analysis of laws governing the distribution of zeros of solutions as well as the asymptotic behavior of solutions of differential equations under consideration.

Oscillation theory has been attracting wide attention of researchers for the past five decades, and as a consequence of their great efforts it has grown up to be a fertile field covering ordinary differential equations, functional differential equations with deviating arguments, partial differential equations with or without functional arguments and difference equations.

It goes without saying that a vast literature, research papers and books, dealing with oscillation theory has been published so far. As regards books and monographs on the subject, most of the existing ones are exclusively concerned with ordinary differential equations and/or difference equations, and there seems to be none which is devoted to the study of oscillations of partial differential equations.

The theory of time scales, which has recently received a lot of attention, was initiated by Hilger [1] in his Ph.D. thesis in 1988 in order to contain both difference and differential calculus in a consistent way. Since then, many authors have expounded on various aspects of the theory of dynamic equations on time scales. For example, the monographes [2–4] and the references cited therein. The present chapter is designed to provide the reader with an exposition of some aspects of the oscillation theory of first order delay partial dynamic equations on time scales.

The chapter is organized as follows. In the next section we give some basic facts of time scale calculus. In Section 3 and Section 4, we make overview of some recent results for oscillations and stability of first order delay dynamic equations on time scales. In Section 5, we investigate a class BVP of first order delay partial dynamic equations on time scales for existence of solutions. In Section 6, we deduct some oscillations criteria for first order partial dynamic equations on time scales.

2. TIME SCALES ESSENTIALS

Definition 1. *A time scale is an arbitrary nonempty closed subset of the real numbers.*

We will denote a time scale by the symbol $\mathbb{T}$.

Definition 2. *For $t \in \mathbb{T}$, define the forward jump operator $\sigma : \mathbb{T} \to \mathbb{T}$ as follows*
$$\sigma(t) = \inf\{s \in \mathbb{T} : s > t\}.$$

Note that $\sigma(t) \geq t$ for any $t \in \mathbb{T}$.

Definition 3. *For $t \in \mathbb{T}$, define the graininess function $\mu : \mathbb{T} \to [0, \infty)$ as follows*
$$\mu(t) = \sigma(t) - t.$$

Definition 4. *For $t \in \mathbb{T}$, define the backward jump operator $\rho : \mathbb{T} \to \mathbb{T}$ by*
$$\rho(t) = \sup\{s \in \mathbb{T} : s < t\}.$$

Observe that $\rho(t) \leq t$ for any $t \in \mathbb{T}$.

Definition 5. *We set*
$$\inf \emptyset = \sup \mathbb{T}, \quad \sup \emptyset = \inf \mathbb{T}.$$

Definition 6. *We define the set*
$$\mathbb{T}^\kappa = \begin{cases} \mathbb{T} \setminus (\rho(\sup \mathbb{T}), \sup \mathbb{T}] & \text{if} \quad \sup \mathbb{T} < \infty \\ \mathbb{T} & \text{otherwise.} \end{cases}$$

Definition 7. *For $t \in \mathbb{T}$, we have the following cases.*

1. *If $\sigma(t) > t$, then we say that t is right-scattered.*

2. *If $t < \sup \mathbb{T}$ and $\sigma(t) = t$, then we say that t is right-dense.*

3. *If $\rho(t) < t$, then we say that t is left-scattered.*

4. *If $t > \inf \mathbb{T}$ and $\rho(t) = t$, then we say that t is left-dense.*

5. *If t is left-scattered and right-scattered at the same time, then we say that t is isolated.*

6. *If t is left-dense and right-dense at the same time, then we say that t is dense.*

Definition 8. *Assume that $f : \mathbb{T} \to \mathbb{R}$ is a function and let $t \in \mathbb{T}^\kappa$. We define $f^\Delta(t)$ to be the number, provided it exists, as follows: for any $\epsilon > 0$ there is a neighbourhood U of t, $U = (t - \delta, t + \delta) \cap \mathbb{T}$ for some $\delta > 0$, such that*

$$|f(\sigma(t)) - f(s) - f^\Delta(t)(\sigma(t) - s)| \leq \epsilon |\sigma(t) - s| \quad \text{for all} \quad s \in U.$$

We say $f^\Delta(t)$ the delta or Hilger derivative of f at t.
We say that f is delta or Hilger differentiable, shortly differentiable, in $\mathbb{T}^\kappa$ if $f^\Delta(t)$ exists for all $t \in \mathbb{T}^\kappa$. The function $f^\Delta : \mathbb{T} \to \mathbb{R}$ is said to be delta derivative or Hilger derivative, shortly derivative, of f in $\mathbb{T}^\kappa$.

Remark 1. *If $\mathbb{T} = \mathbb{R}$, then the delta derivative coincides with the classical derivative.*

Note that the delta derivative is well-defined.

Theorem 1 ([2]- [4]). *Assume $f : \mathbb{T} \to \mathbb{R}$ is a function and let $t \in \mathbb{T}^\kappa$. Then we have the following.*

1. *If f is delta-differentiable at t, then f is continuous at t.*

2. *If f is continuous at t and t is right-scattered, then f is delta-differentiable at t with*

$$f^\Delta(t) = \frac{f(\sigma(t)) - f(t)}{\mu(t)}.$$

3. If t is right-dense, then f is delta-differentiable iff the limit

$$\lim_{s \to t} \frac{f(t) - f(s)}{t - s}$$

exists as a finite number. In this case,

$$f^\Delta(t) = \lim_{s \to t} \frac{f(t) - f(s)}{t - s}.$$

4. If f is delta-differentiable at t, then

$$f(\sigma(t)) = f^\sigma(t) = f(t) + \mu(t) f^\Delta(t).$$

5. Assume f, g are delta-differentiable at t. Then

$$(fg)^\Delta(t) = f^\Delta(t) g(t) + f^\sigma(t) g^\Delta(t) = f(t) g^\Delta(t) + f^\Delta(t) g^\sigma(t)$$

$$\left(\frac{f}{g}\right)^\Delta(t) = \frac{f^\Delta(t) g(t) - f(t) g^\Delta(t)}{g(t) g^\sigma(t)},$$

provided that $g(t), g^\sigma(t) \neq 0$. The product rule gives for $m \in \mathbb{N}$ and $a \in \mathbb{T}$

$$f^\Delta(t) = \sum_{j=0}^{m-1} (\sigma(t) - a)^j (t - a)^{m-j-1},$$

provided that $f(t) = (t-a)^m$. Particularly, $f^\Delta(t) = \sum_{j=0}^{m-1} \sigma^j(t) t^{m-j-1}$

holds, if $f(t) = t^m$.

Definition 9. *A function $f : \mathbb{T} \to \mathbb{R}$ is called regulated, if its right-sided limits exist(finite) at all right-dense points in $\mathbb{T}$ and its left-sided limits exist(finite) at all left-dense points in $\mathbb{T}$.*

Definition 10. *A continuous function $f : \mathbb{T} \to \mathbb{R}$ is called pre-differentiable with region of differentiation D, provided*

1. $D \subset \mathbb{T}^\kappa$,

2. $\mathbb{T}^\kappa \setminus D$ *is countable and contains no right-scattered elements of $\mathbb{T}$,*

3. f is differentiable at each $t \in D$.

Theorem 2 ([2]- [4]). *Let $t_0 \in \mathbb{T}$, $x_0 \in \mathbb{R}$, $f : \mathbb{T}^\kappa \to \mathbb{R}$ be a given regulated map. Then there exists exactly one pre-differentiable function F satisfying*
$$F^\Delta(t) = f(t) \quad \text{for all} \quad t \in D, \quad F(t_0) = x_0.$$

Definition 11. *Assume $f : \mathbb{T} \to \mathbb{R}$ is a regulated function. Any function F by Theorem 2 is called a pre-antiderivative of f. We define the indefinite integral of a regulated function f by*
$$\int f(t)\Delta t = F(t) + c,$$
where c is an arbitrary constant and F is a pre-antiderivative of f. We define the Cauchy integral by
$$\int_\tau^s f(t)\Delta t = F(s) - F(\tau) \quad \text{for all} \quad \tau, s \in \mathbb{T}.$$
A function $F : \mathbb{T} \to \mathbb{R}$ is called an antiderivative of $f : \mathbb{T} \to \mathbb{R}$ provided
$$F^\Delta(t) = f(t) \quad \text{holds for all} \quad t \in \mathbb{T}^\kappa.$$

Definition 12. *A function $f : \mathbb{T} \to \mathbb{R}$ is called rd-continuous if it is continuous at right-dense points of $\mathbb{T}$ and its left-sided limits exist(finite) at left-dense points of $\mathbb{T}$. The set of rd-continuous functions $f : \mathbb{T} \to \mathbb{R}$ will be denoted by $\mathcal{C}_{rd}(\mathbb{T})$. The set of functions $f : \mathbb{T} \to \mathbb{R}$ that are differentiable on $\mathbb{T}$ and whose derivatives are rd-continuous is denoted by $\mathcal{C}^1_{rd}(\mathbb{T})$.*

Note that if f is rd-continuous, then f is regulated.

Theorem 3 ([2]- [4]). *If $a, b, c \in \mathbb{T}$, $\alpha \in \mathbb{R}$ and $f, g \in \mathcal{C}_{rd}(\mathbb{T})$, then*

(i) $\displaystyle\int_a^b (f(t) + g(t))\Delta t = \int_a^b f(t)\Delta t + \int_a^b g(t)\Delta t,$

(ii) $\displaystyle\int_a^b (\alpha f)(t)\Delta t = \alpha \int_a^b f(t)\Delta t,$

(iii) $\displaystyle\int_a^b f(t)\Delta t = -\int_b^a f(t)\Delta t,$

(iv) $\int_a^b f(t)\Delta t = \int_a^c f(t)\Delta t + \int_c^b f(t)\Delta t,$

(v) $\int_a^b f(\sigma(t))g^\Delta(t)\Delta t = (fg)(b) - (fg)(a) - \int_a^b f^\Delta(t)g(t)\Delta t,$

(vi) $\int_a^b f(t)g^\Delta(t)\Delta t = (fg)(b) - (fg)(a) - \int_a^b f^\Delta(t)g(\sigma(t))\Delta t,$

(vii) $\int_a^a f(t)\Delta t = 0,$

(viii)
$$\left|\int_a^b f(t)\Delta t\right| \leq \int_a^b g(t)\Delta t,$$
provided that $|f(t)| \leq g(t), t \in [a,b].$

(ix) $\int_a^b f(t)\Delta t \geq 0,$ *provided that* $f(t) \geq 0, t \in [a,b].$

Definition 13. *We say that* $f: \mathbb{T} \to \mathbb{R}$ *is regressive provided*
$$1 + \mu(t)f(t) \neq 0, \quad t \in \mathbb{T}.$$

We denote by $\mathcal{R}$ *the set of all regressive and rd-continuous functions. Define*
$$\mathcal{R}_+ = \{f \in \mathcal{R} : 1 + \mu(t)f(t) > 0, \quad t \in \mathbb{T}\}.$$

Definition 14. *If* $f, g \in \mathcal{R}$, *then we define*
$$f \oplus g = f + g + \mu f g, \quad \ominus g = -\frac{g}{1 + \mu g}, \quad f \ominus g = f \oplus (\ominus g).$$

Definition 15. *If* $f: \mathbb{T} \to \mathbb{R}$ *is rd-continuous and regressive, then the exponential function* $e_f(\cdot, t_0)$ *is for each fixed* $t_0 \in \mathbb{T}$ *the unique solution of the initial value problem*
$$x^\Delta = f(t)x \quad on \quad \mathbb{T}, \quad x(t_0) = 1.$$

For properties of regressive functions, rd-continuous functions and the exponential function we refer the reader to [2]- [4]. Let $n \in \mathbb{N}$ be fixed. For each $i \in \{1, \ldots, n\}$, we denote by $\mathbb{T}_i$ a time scale.

Definition 16. *The set*

$$\Lambda^n = \mathbb{T}_1 \times \cdots \times \mathbb{T}_n = \{t = (t_1, \ldots, t_n) : t_i \in \mathbb{T}_i, \, i \in \{1, \ldots, n\}\}$$

is called an n-dimensional time scale.

Definition 17. *Let σ_i, $i \in \{1, \ldots, n\}$, be the forward jump operator of $\mathbb{T}_i$. The operator $\sigma : \Lambda^n \to \mathbb{R}^n$, defined by*

$$\sigma(t) = (\sigma_1(t_1), \ldots, \sigma_n(t_n)), \quad t_i \in \mathbb{T}_i, \quad i \in \{1, \ldots, n\},$$

$t = (t_1, \ldots, t_n)$, *is said to be the forward jump operator of Λ^n.*

Definition 18. *Let ρ_i, $i \in \{1, \ldots, n\}$, be the backward jump operator of $\mathbb{T}_i$. The operator $\rho : \Lambda^n \to \mathbb{R}^n$, defined by*

$$\rho(t) = (\rho_1(t_1), \ldots, \rho_n(t_n)), \quad t_i \in \mathbb{T}_i, \quad i \in \{1, \ldots, n\},$$

is said to be the backward jump operator of Λ^n.

Definition 19. *For $x = (x_1, \ldots, x_n)$, $y = (y_1, \ldots, y_n) \in \mathbb{R}^n$, we write $x \geq y$ whenever $x_i \geq y_i$ for all $i \in \{1, \ldots, n\}$. In a similar way, we understand $x > y$, $x < y$ and $x \leq y$.*

Definition 20. *The graininess function $\mu : \Lambda^n \to [0, \infty)^n$ is defined by*

$$\mu(t) = (\mu_1(t_1), \ldots, \mu_n(t_n)), \quad t_i \in \mathbb{T}_i, \quad i \in \{1, \ldots, n\},$$

$t = (t_1, \ldots, t_n)$.

Definition 21. *Let $f : \Lambda^n \to \mathbb{R}$. We introduce the following notations.*

$$f^\sigma(t) = f(\sigma_1(t_1), \ldots, \sigma_n(t_n)),$$

$$f_i^{\sigma_i}(t) = f(t_1, \ldots, t_{i-1}, \sigma_i(t_i), t_{i+1}, \ldots, t_n),$$

$$f_{i_1 \ldots i_l}^{\sigma_{i_1} \ldots \sigma_{i_l}}(t) = f(\ldots, \sigma_{i_1}(t_{i_1}), \ldots, \sigma_{i_2}(t_{i_2}), \ldots, \sigma_{i_l}(t_{i_l}), \ldots),$$

where $1 \leq i_1 < \ldots < i_l \leq n$, $i_m \in \mathbb{N}$, $m \in \{1, \ldots, l\}$, $t_i \in \mathbb{T}_i$, $t_{i_m} \in \mathbb{T}_{i_m}$, $i \in \{1, \ldots, n\}$, $m \in \{1, \ldots, l\}$, $l \in \mathbb{N}$.

Definition 22. *We set*

$$\Lambda^{\kappa n} = \mathbb{T}_1^\kappa \times \ldots \times \mathbb{T}_n^\kappa,$$

$$\Lambda_i^{\kappa_i n} = \mathbb{T}_1 \times \ldots \times \mathbb{T}_{i-1} \times \mathbb{T}_i^\kappa \times \mathbb{T}_{i+1} \times \ldots \times \mathbb{T}_n, \quad i \in \{1, \ldots, n\},$$

$$\Lambda_{i_1 \ldots i_l}^{\kappa_{i_1} \ldots \kappa_{i_l} n} = \ldots \times \mathbb{T}_{i_1}^\kappa \times \ldots \times \mathbb{T}_{i_2}^\kappa \times \ldots \times \mathbb{T}_{i_l}^\kappa \times \ldots,$$

where $1 \leq i_1 < \ldots < i_l \leq n$, $i_m \in \mathbb{N}$, $m \in \{1, 2, \ldots, l\}$.

Remark 2. *If* $(i_1, \ldots, i_l) = (1, 2, \ldots, n)$, *then*

$$\Lambda_{i_1 \ldots i_l}^{\kappa_1 \ldots \kappa_l n} = \Lambda^{\kappa n}.$$

Definition 23. *Assume that* $f : \Lambda^n \to \mathbb{R}$ *is a function and let* $t \in \Lambda_i^{\kappa_i n}$. *We define*

$$\frac{\partial f}{\Delta_i t_i}(t) = f_{t_i}^{\Delta_i}(t)$$

to be the number, provided it exists, with the property that for any $\varepsilon_i > 0$, *there exists a neighbourhood*

$$U_i = (t_i - \delta_i, t_i + \delta_i) \cap \mathbb{T}_i,$$

for some $\delta_i > 0$, *such that*

$$\left| f(t_1, \ldots, t_{i-1}, \sigma_i(t_i), t_{i+1}, \ldots, t_n) - f(t_1, \ldots, t_{i-1}, s_i, t_{i+1}, \ldots, t_n) \right.$$
$$\left. - f_{t_i}^{\Delta_i}(t)(\sigma_i(t_i) - s_i) \right| \leq \varepsilon_i |\sigma_i(t_i) - s_i| \quad \textit{for all} \quad s_i \in U_i.$$

We call $f_{t_i}^{\Delta_i}(t)$ *the partial delta derivative (or partial Hilger derivative) of* f *with respect to* t_i *at* t. *We say that* f *is partial delta differentiable (or partial Hilger differentiable) with respect to* t_i *in* $\Lambda_i^{\kappa_i n}$ *if* $f_{t_i}^{\Delta_i}(t)$ *exists for all* $t \in \Lambda_i^{\kappa_i n}$. *The function* $f_{t_i}^{\Delta_i} : \Lambda_i^{\kappa_i n} \to \mathbb{R}$ *is said to be the partial delta derivative (or partial Hilger derivative) with respect to* t_i *of* f *in* $\Lambda_i^{\kappa_i n}$.

The partial delta derivative is well defined. For the properties of the partial delta derivative we refer the reader to [4].

Definition 24. *For a function $f : \Lambda^n \to \mathbb{R}$, we shall talk about the second-order partial delta derivative with respect to t_i and t_j, $i, j \in \{1, \ldots, n\}$, $f_{t_i t_j}^{\Delta_i \Delta_j}$, provided $f_{t_i}^{\Delta_i}$ is partial delta differentiable with respect to t_j on $\Lambda_{ij}^{\kappa_i \kappa_j n} = (\Lambda_i^{\kappa_i n})_j^{\kappa_j n}$ with partial delta derivative*

$$f_{t_i t_j}^{\Delta_i \Delta_j} = \left(f_{t_i}^{\Delta_i}\right)_{t_j}^{\Delta_j} : \Lambda_{ij}^{\kappa_i \kappa_j n} \to \mathbb{R}.$$

For $i = j$, we will write

$$f_{t_i t_i}^{\Delta_i \Delta_i} = f_{t_i}^{\Delta_i^2}.$$

Similarly, we define higher order partial delta derivatives

$$f_{t_i t_j \ldots t_l}^{\Delta_i \Delta_j \ldots \Delta_l} : \Lambda_{ij\ldots l}^{\kappa_i \kappa_j \ldots \kappa_l n} \to \mathbb{R}.$$

For $t \in \Lambda^n$, we define

$$\sigma^2(t) = \sigma(\sigma(t)) = (\sigma_1(\sigma_1(t_1)), \sigma_2(\sigma_2(t_2)), \ldots, \sigma_n(\sigma_n(t_n))).$$

Now we will introduce the conception for multiple integration on time scales. Suppose $a_i < b_i$ are points in $\mathbb{T}_i$ and $[a_i, b_i)$ is the half-closed bounded interval in $\mathbb{T}_i$, $i \in \{1, \ldots, n\}$. Let us introduce a "rectangle" in $\Lambda^n = \mathbb{T}_1 \times \ldots \times \mathbb{T}_n$ by

$$\begin{aligned}R &= [a_1, b_1) \times \ldots \times [a_n, b_n) \\ &= \{(t_1, \ldots, t_n) : t_i \in [a_i, b_i), i \in \{1, \ldots, n\}\}.\end{aligned}$$

Let

$$a_i = t_i^0 < t_i^1 < \ldots < t_i^{k_i} = b_i.$$

Definition 25. *We call the collection of intervals*

$$P_i = \left\{[t_i^{j_i-1}, t_i^{j_i}) : j_i \in \{1, \ldots, k_i\}\right\}, \quad i \in \{1, \ldots, n\},$$

a Δ_i-partition of $[a_i, b_i)$ and denote the set of all Δ_i-partitions of $[a_i, b_i)$ by $\mathcal{P}_i([a_i, b_i))$.

Definition 26. *Let*

$$R_{j_1\ldots j_n} = [t_1^{j_1-1}, t_1^{j_1}) \times \ldots \times [t_n^{j_n-1}, t_n^{j_n})$$

$$1 \leq j_i \leq k_i, \quad i \in \{1, \ldots, n\}.$$

We call the collection

$$P = \{R_{j_1\ldots j_n} : 1 \leq j_i \leq k_i,\ i \in \{1, \ldots, n\}\}$$

a Δ-partition of R, generated by the Δ_i-partitions P_i of $[a_i, b_i)$, and we write

$$P = P_1 \times \ldots \times P_n.$$

The set of all Δ-partitions of R is denoted by $\mathcal{P}(R)$. Moreover, for a bounded function $f : R \to \mathbb{R}$, we set

$$M = \sup\{f(t_1, \ldots, t_n) : (t_1, \ldots, t_n) \in R\},$$

$$m = \inf\{f(t_1, \ldots, t_n) : (t_1, \ldots, t_n) \in R\},$$

$$M_{j_1\ldots j_n} = \sup\{f(t_1, \ldots, t_n) : (t_1, \ldots, t_n) \in R_{j_1\ldots j_n}\},$$

$$m_{j_1\ldots j_n} = \inf\{f(t_1, \ldots, t_n) : (t_1, \ldots, t_n) \in R_{j_1 j_2\ldots j_n}\}.$$

Definition 27. *The upper Darboux Δ-sum $U(f, P)$ and the lower Darboux Δ-sum $L(f, P)$ with respect to P are defined by*

$$U(f, P) = \sum_{j_1=1}^{k_1} \sum_{j_2=1}^{k_2} \ldots \sum_{j_n=1}^{k_n} M_{j_1 j_2 \ldots j_n}(t_1^{j_1} - t_1^{j_1-1})(t_2^{j_2} - t_2^{j_2-1}) \ldots (t_n^{j_n} - t_n^{j_n-1})$$

and

$$L(f, P) = \sum_{j_1=1}^{k_1} \sum_{j_2=1}^{k_2} \ldots \sum_{j_n=1}^{k_n} m_{j_1 j_2 \ldots j_n}(t_1^{j_1} - t_1^{j_1-1})(t_2^{j_2} - t_2^{j_2-1}) \ldots (t_n^{j_n} - t_n^{j_n-1}).$$

Definition 28. *The upper Darboux Δ-integral $U(f)$ of f over R and the lower Darboux Δ-integral $L(f)$ of f over R are defined by*

$$U(f) = \inf\{U(f, P) : P \in \mathcal{P}(R)\} \quad \text{and} \quad L(f) = \sup\{L(f, P) : P \in \mathcal{P}(R)\}.$$

We have that $U(f)$ and $L(f)$ are finite real numbers.

Definition 29. *We say that f is Δ-integrable over R provided $L(f) = U(f)$. In this case, we write*

$$\int_R f(t_1, \ldots, t_n) \Delta_1 t_1 \Delta_2 t_2 \ldots \Delta_n t_n$$

for this common value. We call this integral the Darboux Δ-integral.

For the properties of the multiple delta integral we refer the reader to [4].

3. Oscillations and Nonoscillations Tests for First Order DelayDynamic Equations

In this section we will list some well-known results on the oscillations and and nonoscillations of delay dynamic equations. Suppose that $\mathbb{T}$ is a unbounded above time scale with forward jump operator and delta differentiation operator σ and Δ, respectively. Consider the delay dynamic equation

$$x^\Delta(t) + p(t) x(\tau(t)) = 0, \quad t \geq t_0, \tag{1}$$

where

(A1) $p \in C_{rd}([t_0, \infty))$, $p(t) \geq 0$, $t \in [t_0, \infty)$,

(A2) $\tau \in C([t_0, \infty))$, $\tau : [t_0, \infty) \to \mathbb{T}$, $\tau(t) \leq t$ for all large t and $\lim_{t \to \infty} \tau(t) = \infty$.

As a customary, the equation (1) is called oscillatory provided that it does not possess any eventually positive(negative) solutions.

Theorem 4 ([5]). *(Zhang and all, 2002) Let*

$$E = \{\lambda \geq 0 : 1 - \lambda p(t) > 0 \quad \text{for all large } t\}.$$

Assume $(A1)$, $(A2)$ *and*

$$\liminf_{t \to \infty} \inf_{\lambda \in E} \left(\frac{1}{\lambda e_{-\lambda p}(\tau(t), t)} \right) > 1. \tag{2}$$

Then the equation (1) is oscillatory.

Due to the results in [6](Bohner, 2005) and [7](Bohner and all, 2008), the condition (2) can be restated as

$$\liminf_{t\to\infty} \inf_{\substack{-\lambda p \in \mathcal{R}^+([\tau(t),t)) \\ \lambda \geq 0}} \left(\frac{1}{\lambda e_{-\lambda p}(\tau(t),t)}\right) > 1,$$

where

$$\mathcal{R}^+(J) = \{f \in \mathcal{C}_{rd}(J) : 1 + \mu(t)p(t) > 0, \quad t \in J\}$$

and J is an interval.

Theorem 5 ([5]). *(Zhang and all, 2002) Suppose $(A1)$ and $(A2)$. Assume that there exist $s \in [t_0, \infty)$ and $\lambda_0 \geq 0$ such that $-\lambda_0 p \in \mathcal{R}^+([s, \infty))$ and*

$$\frac{1}{\lambda_0 e_{-\lambda_0 p}(\tau(t),t)} \leq 1$$

for all $t \in [s, \infty)$. Then the equation (1) has a nonoscillatory solution.

Note that Theorem 4 and Theorem 5 use integration over the interval $[\tau(t), t)$. Now we will quote some results which involve integration over the interval $[\tau(t), t]$.

Theorem 6 ([8]). *(Braverman and all, 2010) Suppose $(A1)$ and $(A2)$. Assume that τ is eventually nondecreasing and*

$$\limsup_{t\to\infty} \int_{\tau(t)}^{\sigma(t)} p(\eta)\Delta\eta > 1.$$

Then the equation (1) is oscillatory.

The next two results depend on integration over the interval $[\tau(t), t)$.

Theorem 7 ([8]). *(Braverman and all, 2010) Suppose $(A1)$ and $(A2)$. Assume that τ is eventually nondecreasing and there exists an $\alpha \in (0, 1)$ such that*

$$\liminf_{t\to\infty} \int_{\tau(t)}^{t} p(\eta)\Delta\eta > \alpha \quad \text{and} \quad \limsup_{t\to\infty} \int_{\tau(t)}^{t} p(\eta)\Delta\eta > 1 - \frac{\alpha^2}{4}.$$

Then the equation (1) is oscillatory.

The next result improve Theorem 7 by repleacing the right-hand side of the second condition by a greater constant.

Theorem 8 ([9]). *(Sahiner and all, 2006) Suppose $(A1)$ and $(A2)$. Assume that τ is eventually nondecreasing and there exists an $\alpha \in (0, 1)$ such that*

$$\liminf_{t\to\infty} \int_{\tau(t)}^{t} p(\eta)\Delta\eta > \alpha \quad \text{and} \quad \limsup_{t\to\infty} \int_{\tau(t)}^{t} p(\eta)\Delta\eta > 1-\left(1-\sqrt{1-\alpha}\right)^2.$$

Then the equation (1) is oscillatory.

The next result improves Theorem 8 to the larger interval $[\tau(t), t]$.

Theorem 9 ([10]). *(Agarwal and all, 2009) Suppose $(A1)$ and $(A2)$. Assume that τ is eventually nondecreasing and there exists an $\alpha \in (0, 1)$ such that*

$$\liminf_{t\to\infty} \int_{\tau(t)}^{t} p(\eta)\Delta\eta > \alpha \quad \text{and} \quad \limsup_{t\to\infty} \int_{\tau(t)}^{\sigma(t)} p(\eta)\Delta\eta > 1-\left(1-\sqrt{1-\alpha}\right)^2.$$

Then the equation (1) is oscillatory.

Now we will give an other kind oscillation test for the equation (1).

Theorem 10 ([11]). *(Karpuz, 2019) Suppose $(A1)$ and $(A2)$. Assume*

$$\liminf_{t\to\infty} \inf_{\lambda\in[1,\infty)} \left(\frac{1}{\lambda} e_{\lambda p}(\sigma(t), \tau(t))\right) > 1.$$

Then the equation (1) is oscillatory.

Next, we have the following nonoscillation test.

Theorem 11 ([11]). *(Karpuz, 2019) Suppose $(A1)$ and $(A2)$. Assume that there exists an $\lambda_0 \in [1, \infty)$ such that*

$$\frac{1}{\lambda_0} e_{\lambda_0 p}(\sigma(t), \tau(t)) \leq 1$$

for all large $t \in [t_0, \infty)$. Then the equation (1) is nonoscillatory.

Now we consider the dynamic equation

$$x^{\Delta}(t) + \sum_{i=1}^{n} p_i(t) x(\tau_i(t)) = 0, \qquad (3)$$

where

(B1) $p_i \in C_{rd}(\mathbb{T}), p_i \geq 0$ on $\mathbb{T}, i \in \{1, \ldots, n\}$,

(B2) $\tau_i : \mathbb{T} \to \mathbb{T}, \tau_i(t) < t, t \in \mathbb{T}, i \in \{1, \ldots, n\}$.

The next result gives an oscillation criterion for the equation (3).

Theorem 12 ([12]). *(Agwo, 2008) Suppose (B1) and (B2). If*

$$\limsup_{t_0 \to \infty} \sup_{t > t_0} \sup_{\lambda \in E} \frac{\lambda \sum_{i=1}^{n} p_i(t)}{\sum_{i=1}^{n} p_i(t) \exp\left(-\int_{\tau_i(t)}^{t} \frac{1}{\mu(s)} \mathrm{Log}\left(1 - \lambda \mu(s) \sum_{i=1}^{n} p_i(s)\right) \Delta s\right)} < 1,$$

where

$$E = \left\{ \lambda > 0 : 1 - \lambda \left(\sum_{i=1}^{n} p_i(t) \right) \mu(t) > 0 \right\},$$

then the equation (3) is oscillatory.

4. STABILITY OF FIRST ORDER DYNAMIC EQUATIONS

In this section we will list some criteria for uniform exponential stability and global stability of some classes first order dynamic equations. Consider the equation

$$x^{\Delta}(t) + \sum_{i=1}^{n} a_i(t) x(\tau_i(t)) = 0, \quad t \geq t_0,$$

$$x(t_0) = x_0, \quad x(t) = \phi(t), \quad t \in [t_{-1}, t_0),$$
(4)

where

(C1) $x_0 \in \mathbb{R}, \phi \in C_{rd}([t_{-1}, t_0))$, ϕ has a finite left-sided limit at the initial point provided that it is left-dense.

(C2) $\tau_i : \mathbb{T} \to \mathbb{T}, \tau_i(t) \leq t, \lim_{t \to \infty} \tau_i(t) = \infty, i \in \{1, \ldots, n\}$.

Here $t_{-1} = \inf_{t \in [t_0, \infty)} \{\min_{i \in \{1, \ldots, n\}} \tau_i(t)\}$.

Definition 30. *The trivial solution of the equation (4) is said to be uniformly exponentially stable if there exist positive constants M and λ such that for any $s \in [t_0, \infty)$, the solution x of the associated homogeneous IVP*

$$x^{\Delta}(t) + \sum_{i=1}^{n} a_i(t) x(\tau_i(t)) = 0, \quad t \geq s,$$

$$x(t_0) = x_0, \quad x(t) = \phi(t), \quad t \in [s_{-1}, s),$$

where $s_{-1} = \inf_{t \in [s,\infty)} \{\min_{i \in \{1,\ldots,n\}} \tau_i(t)\}$, satisfies

$$|x(t)| \leq M e_{\ominus \lambda}(t,s)\left(|x_0| + \sup_{\eta \in [s_{-1},s)} |\phi(\eta)|\right), \quad t \in [s,\infty).$$

With $BC_{rd}([t_0,\infty))$ we will denote the set of all bounded rd-continuous functions. Below we will list some criteria for uniformly exponential stability of the equation (4).

Theorem 13 ([13]). *(Braverman and all, 2012) Let $a_i \geq 0$ on $[t_0,\infty)$, $a_i \in BC_{rd}([t_0,\infty))$, $i \in \{1,\ldots,n\}$,*

$$0 < \liminf_{t \to \infty} \sum_{i=1}^{n} a_i(t) \leq \limsup_{t \to \infty} \sum_{i=1}^{n} a_i(t) < \infty \tag{5}$$

and

$$\limsup_{t \to \infty} \sum_{i=1}^{n} a_i(t) \int_{\tau_i(t)}^{\sigma(t)} \sum_{j=1}^{n} a_j(\eta)\Delta\eta < \liminf_{t \to \infty} \sum_{i=1}^{n} a_i(t).$$

Then the trivial solution of the equation (4) is uniformly exponential stable.

The above result involves integration over $[\tau_i(t),t]$, $i \in \{1,\ldots,,n\}$. The next result involves integration over $[\tau_i(t),t)$, $i \in \{1,\ldots,,n\}$.

Theorem 14 ([13]). *(Braverman and all, 2012) Let $a_i \geq 0$ on $[t_0,\infty)$, $a_i \in BC_{rd}([t_0,\infty))$, $\tau_i(\sigma(t)) \leq t$, $t \in [t_0,\infty)$, $i \in \{1,\ldots,n\}$, (5) and*

$$\limsup_{t \to \infty} \sum_{i=1}^{n} a_i(t) \int_{\tau_i(t)}^{t} \sum_{j=1}^{n} a_j(\eta)\Delta\eta < \liminf_{t \to \infty} \sum_{i=1}^{n} a_i(t).$$

Then the trivial solution of the equation (4) is uniformly exponential stable.

Now we consider the nonlinear delay dynamic equation

$$x^{\Delta}(t) = -\sum_{j=0}^{n} a_j(t)f_j(x(\tau_j(t))), \quad t \in [t_0,\infty),$$
$$x(t) = \psi(t), \quad t \in [\tau_n(t_0),t_0], \tag{6}$$

where

(D1) $\tau_j : \mathbb{T} \to \mathbb{T}, j \in \{0, \ldots, n\}$, are right-dense strictly increasing functions unbounded above with $\tau_n(\mathbb{T}) = \mathbb{T}$ such that there is a constant M for which
$$\rho^n(t) - M \leq \tau_n(t) < \tau_{n-1}(t) < \ldots < \tau_0(t) = t, \quad j \in \{0, \ldots, n\},$$

(D2) $f_j : \mathbb{R} \to \mathbb{R}$, there exists a strictly increasing continuous function f on $(-\infty, \infty)$ such that
$$f(0) = 0, \quad \frac{f_j(x)}{f(x)} \geq 1, \quad x \neq 0, \quad j \in \{0, \ldots, n\},$$
$\lim_{x \to \infty} f(x)$ is finite if $f(x) \neq x$,

(D3) $a_j \in \mathcal{C}_{rd}(\mathbb{T})$,
$$a_j(t) \geq 0, \quad \sum_{j=0}^{n} a_j(t) > 0, \quad t \in [t_0, \infty),$$
$$\int_{t_0}^{\infty} \sum_{j=0}^{n} a_j(t) \Delta t = \infty.$$

Definition 31. *The zero solution of the equation (6) is said to be uniformly stable if for any $\epsilon > 0$ and any $t^* \in [t_0, \infty)$ there exists a $\delta = \delta(\epsilon) > 0$ such that*
$$\sup_{s \in [\tau_n(t^*), \sigma(t^*)]} |x(s)| < \delta$$
implies that the solution x of the equation (6) satisfies $|x(t)| < \epsilon$ for $t \in [t^*, \infty)$.

Let
$$r_1 = \sup_{s \geq \tau_n^{-1}(t_0)} \sum_{j=0}^{n} \int_{\tau_j^{-1}(\tau_n(s))}^{\sigma(s)} a_j(t) \Delta t,$$

$$r_2 = \sup_{s \geq \tau_n^{-1}(t_0)} \sum_{j=1}^{n} \int_{\tau_n(s)}^{\tau_j^{-1}(\tau_n(s))} a_j(t) \Delta t,$$

$$\phi(x) = x - r_1 f(x), \quad x \in \mathbb{R}.$$

We have the following criterion for uniform stability of the zero solution of the equation (6).

Theorem 15 ([14]). *(Anderson, 2008) Assume* (D1)-(D3), $f(x) \neq x$,

$$\sum_{t \geq \tau_n^{-1}(t_0)} \int_{\tau_n(t)}^{\sigma(t)} \sum_{j=0}^{n} a_j(s) \Delta s < \infty,$$

ϕ *is monotone on* $(-\infty, \infty)$ *and for any* $L < 0$ *we have*

$$-r_2 f(-r_2 f(L)) > L$$

if ϕ *is increasing, and*

$$\phi(\phi(L) - r_2 f(L)) - r_2 f(\phi(L) - r_2 f(L)) > L$$

if ϕ *is decreasing. Then the zero solution of the equation* (6) *is uniformly stable.*

5. EXISTENCE OF SOLUTIONS OF IBVPS FOR A CLASS FIRST ORDER DELAY PARIAL DYNAMIC EQUATIONS

Suppose that $\mathbb{T}_1$ is an unbounded above time scale and $\mathbb{T}_2$ is a time scale with forward jump operators and delta differentiation operators σ_1, σ_2 and Δ_1, Δ_2, respectively. Let $t_1^0 \in \mathbb{T}_1$ and $a, b \in \mathbb{T}_2$, $a < b$. Consider the IBVP for the following first order partial dynamic equation.

$$u_{t_1}^{\Delta_1}(t_1, t_2) + u_{t_2}^{\Delta_2}(\tau_0(t_1), t_2) + \sum_{j=0}^{n} b_j(t_1) u(\tau_j(t_1), t_2)$$

$$= f(t_1, t_2, u(t_1, t_2), u(\tau_0(t_1), t_2), u(\tau_1(t_1), t_2), \ldots, u(\tau_n(t_1), t_2)), \quad (7)$$

$$t_1 \geq t_1^0, \quad t_2 \in [a, b],$$

$$u(t_1, a) = u(t_1, b), \quad t_1 \in [t_1^{-1}, \infty), \quad (8)$$

$$u(t_1, t_2) = \phi(t_1, t_2), \quad t_1 \in [t_1^{-1}, t_1^0], \quad t_2 \in [a, b], \quad (9)$$

where

(G1) $b_j \in \mathcal{C}_{rd}([t_1^0, \infty))$, $j \in \{0, 1, \ldots, n\}$,

(G2) $\tau_j : [t_1^0, \infty) \to \mathbb{T}$, $\tau_j \in C_{rd}([t_1^0, \infty))$, $\tau_j(t_1) \leq t_1$, $t_1 \in \mathbb{T}_1$, $\lim_{t_1 \to \infty} \tau_j(t_1) = \infty$, $j \in \{0, \ldots, n\}$,

$$\tau_{min}(t_1) = \min_{j \in \{0,1,\ldots,n\}} \{\tau_j(t_1)\}, \quad t_1 \in [t_1^0, \infty),$$

$$t_1^{-1} = \inf_{t_1 \in [t_1^0, \infty)} \tau_{min}(t_1),$$

$\tau_0 : [t_1^0, \infty) \to [t_1^{-1}, \infty)$ is onto,

(G3) $f : \mathbb{T}_1 \times \mathbb{T}_2 \times \mathbb{R}^{n+2} \to \mathbb{R}$,

$$|f(t_1, t_2, y, x_0, x_1, \ldots, x_n)| \leq c(t_1, t_2)|y|^l + \sum_{j=0}^{n} c_j(t_1, t_2)|x_j|^{l_j},$$

$(t_1, t_2) \in \mathbb{T}_1 \times \mathbb{T}_2$, $(y, x_0, x_1, \ldots, x_n) \in \mathbb{R}^{n+2}$, $c, c_j \in C_{rd}(\mathbb{T}_1 \times \mathbb{T}_2)$, $c, c_j \geq 0$ on $\mathbb{T}_1 \times \mathbb{T}_2$, $l, l_j \in [0, \infty)$, $j \in \{0, 1, \ldots, n\}$,

(G4) $\phi \in C_{rd}\left([t_1^{-1}, \infty) \times [a, b]\right)$.

To prove our existence result, we will use the following fixed point theorems.

Definition 32. *Let (X, d) be a metric space and M be a subset of X. The mapping $T : M \to X$ is said to be expansive if there exists a constant $h > 1$ such that*

$$d(Tx, Ty) \geq h d(x, y)$$

for any $x, y \in M$.

Theorem 16 ([15], Theorem 2.4). *Let X be a nonempty closed convex subset of a Banach space E. Suppose that T and S map X into E such that*

1. *S is continuous and $S(X)$ resides in a compact subset of E.*

2. *$T : X \to E$ is expansive.*

3. *$S(X) \subset (I - T)(E)$ and $[x = Tx + Sy, y \in X] \Longrightarrow x \in X$ (or $S(X) \subset (I - T)(X)$).*

Then there exists a point $x^* \in X$ such that

$$Sx^* + Tx^* = x^*.$$

Theorem 17. *Let X be a nonempty closed convex subset of a Banach space E and Y is a nonempty compact subset of E such that $X \subset Y$, $Y \neq X$. Suppose that T and S map X into E such that*

1. *S is continuous and $S(X)$ resides in Y.*

2. *$T : E \to E$ is linear, continuous and expansive, and T^{-1} is defined on Y and*
$$\{T^{-1}y - z : y \in Y, \quad z \in S(X)\} \subset Y.$$

Then there exists an $x^* \in X$ such that
$$Tx^* + Sx^* = x^*.$$

Proof. Since Y is compact and $S(X)$ resides in Y, we have that the first condition of Theorem 16 holds. Because $T : E \to E$ is expansive and $X \subset E$, we have that the second condition of Theorem 16 holds. Note that $T^{-1} : Y \to E$ exists, it is linear and contractive with a constant $l \in (0, 1)$. Let $z \in S(X)$ be arbitrarily chosen and fixed. Take $y_0 \in Y$ arbitrarily. Define the sequence $\{y_n\}_{n \in \mathbb{N}}$ as follows.
$$y_{n+1} = T^{-1}y_n - z, \quad n \in \mathbb{N}_0.$$

Note that $y_1 = T^{-1}y_0 - z \in Y$. Hence, $y_2 = T^{-1}y_1 - z \in Y$, and so on, $\{y_n\}_{n \in \mathbb{N}} \subset Y$. Then

$$\begin{aligned} \|y_2 - y_1\| &= \|T^{-1}y_1 - T^{-1}y_0\| \\ &\leq l\|y_1 - y_0\|, \\ \|y_3 - y_2\| &= \|T^{-1}y_2 - T^{-1}y_1\| \\ &\leq l\|y_2 - y_1\| \\ &\leq l^2\|y_1 - y_0\|. \end{aligned}$$

Using the principle of the mathematical induction, we get
$$\|y_{n+1} - y_n\| \leq l^n \|y_1 - y_0\|, \quad n \in \mathbb{N}.$$

Now, for $m > n$, $m, n \in \mathbb{N}$, we find

$$\begin{aligned}
\|y_m - y_n\| &\leq \|y_m - y_{m-1}\| + \cdots + \|y_{n+1} - y_n\| \\
&\leq \left(l^{m-1} + \cdots + l^n\right) \|y_1 - y_0\| \\
&\leq l^n \sum_{j=0}^{\infty} l^j \|y_1 - y_0\| \\
&= \frac{l^n}{1-l} \|y_1 - y_0\|.
\end{aligned}$$

Therefore $\{y_n\}_{n\in\mathbb{N}}$ is a Cauchy sequence of elements of $Y \subset E$. Since E is a Banach space, it follows that the sequence $\{y_n\}_{n\in\mathbb{N}}$ is convergent to an element $y^* \in E$. Because $\{y_n\}_{n\in\mathbb{N}} \subset Y$ and $Y \subset E$ is compact, we have that $y^* \in Y$. Thus,

$$y^* = T^{-1}y^* - z$$

or

$$z^* = Tz^* + z, \quad z^* = T^{-1}y^* \in X.$$

Because $z \in S(X)$ was arbitrarily chosen, we conclude that $S(X) \subset (I - T)(X)$, i.e., the third condition of Theorem 16 holds. Hence and Theorem 16, it follows that there exists an $x^* \in X$ such that

$$Tx^* + Sx^* = x^*.$$

This completes the proof. □

To prove our existence result we will use the following useful lemmas.

Lemma 1. *Let $p, q \in C_{rd}^1([a,b])$, $1 + \mu_2 p > 0$ on $[a,b]$ and*

$$G(t_2, s_2) = \begin{cases} e_p(a,b)e_p(t_2, s_2) & \text{if } a < s_2 \leq t_2 < b \\ e_p(t_2, s_2) & \text{if } a < t_2 \leq s_2 < b. \end{cases}$$

Then

$$x(t_2) = \int_a^b G(t_2, s_2) \frac{q(s_2)}{(e_p(a,b) - 1)(1 + \mu_2(s_2)p(s_2))} \Delta_2 s_2, \quad t_2 \in [a,b],$$

is a solution of the BVP

$$x^{\Delta_2}(t_2) = p(t_2)x(t_2) + q(t_2), \quad t_2 \in (a,b), \tag{10}$$

$$x(a) = x(b). \tag{11}$$

Proof. We have

$$\begin{aligned}
x(t_2) &= e_p(a,b)\int_a^{t_2} e_p(t_2,s_2)\frac{q(s_2)}{(e_p(a,b)-1)(1+\mu_2(s_2)p(s_2))}\Delta_2 s_2 \\
&\quad + \int_{t_2}^{b} e_p(t_2,s_2)\frac{q(s_2)}{(e_p(a,b)-1)(1+\mu_2(s_2)p(s_2))}\Delta_2 s_2, \quad t_2 \in [a,b], \\
x(a) &= \int_a^{b} e_p(a,s_2)\frac{q(s_2)}{(e_p(a,b)-1)(1+\mu_2(s_2)p(s_2))}\Delta_2 s_2, \\
x(b) &= e_p(a,b)\int_a^{b} e_p(b,s_2)\frac{q(s_2)}{(e_p(a,b)-1)(1+\mu_2(s_2)p(s_2))}\Delta_2 s_2 \\
&= \int_a^{b} e_p(a,s_2)\frac{q(s_2)}{(e_p(a,b)-1)(1+\mu_2(s_2)p(s_2))}\Delta_2 s_2.
\end{aligned}$$

Hence,
$$x(a) = x(b)$$
and

$$\begin{aligned}
x^{\Delta_2}(t_2) &= e_p(a,b)e_p(\sigma_2(t_2),t_2)\frac{q(t_2)}{(e_p(a,b)-1)(1+\mu_2(t_2)p(t_2))} \\
&\quad + p(t_2)e_p(a,b)\int_a^{t_2} e_p(t_2,s_2)\frac{q(s_2)}{(e_p(a,b)-1)(1+\mu_2(s_2)p(s_2))}\Delta_2 s_2 \\
&\quad - e_p(\sigma_2(t_2),t_2)\frac{q(t_2)}{(e_p(a,b)-1)(1+\mu_2(t_2)p(t_2))} \\
&\quad + p(t_2)\int_{t_2}^{b} e_p(t_2,s_2)\frac{q(s_2)}{(e_p(a,b)-1)(1+\mu_2(s_2)p(s_2))}\Delta_2 s_2 \\
&= (e_p(a,b)-1)e_p(\sigma_2(t_2),t_2)\frac{q(t_2)}{(e_p(a,b)-1)(1+\mu(t_2)p(t_2))} \\
&\quad + p(t_2)x(t_2)
\end{aligned}$$

$$= (1+\mu(t_2)p(t_2))\frac{q(t_2)}{1+\mu(t_2)p(t_2)} + p(t_2)x(t_2)$$

$$= p(t_2)x(t_2) + q(t_2), \quad t_2 \in (a,b).$$

This completes the proof. □

Assume

(G5) $p \in \mathcal{C}_{rd}^1([a,b]), 1+\mu_2 p > 0$ on $[a,b]$.

Lemma 2. *Suppose* $(G1)$-$(G5)$. *Then* $u \in \mathcal{C}_{rd}^1([t_1^0,\infty) \times [a,b])$ *is a solution of the BVP* (7), (8) *if and only if it satisfies the integral equation*

$$\int_{t_1^0}^{t_1} u(\tau_0(s_1), t_2) \Delta_1 s_1 = \int_a^b G(t_2, s_2) \bigg(-p(s_2) \int_{t_1^0}^{t_1} u(\tau_0(s_1), s_2) \Delta_1 s_1$$

$$-u(t_1, s_2) + u(t_1^0, s_2)$$

$$-\sum_{j=0}^n \int_{t_1^0}^{t_1} b_j(s_1) u(\tau_j(s_1), s_2) \Delta_1 s_1$$

$$+ \int_{t_1^0}^{t_1} f(s_1, s_2, u(s_1, s_2), u(\tau_0(s_1), s_2), \ldots, u(\tau_n(s_1), s_2)) \Delta_1 s_1 \bigg) \Delta_2 s_2, \tag{12}$$

$t_1 \in [t_1^0, \infty), t_2 \in [a,b]$.

Proof. 1. Let $u \in \mathcal{C}_{rd}^1([t_1^0,\infty) \times [a,b])$ be a solution of the BVP (7), (8). We integrate both sides of the equation (7) from t_1^0 to t_1 and we get

$$\int_{t_1^0}^{t_1} u_{t_1}^{\Delta_1}(s_1, t_2) \Delta_1 s_1 + \int_{t_1^0}^{t_1} u_{t_2}^{\Delta_2}(\tau_0(s_1), t_2) \Delta_1 s_1$$

$$+ \sum_{j=0}^n \int_{t_1^0}^{t_1} b_j(s_1) u(\tau_j(s_1), t_2) \Delta_1 s_1$$

$$= \int_{t_1^0}^{t_1} f(s_1, t_2, u(s_1, t_2), u(\tau_0(s_1), t_2), \ldots, u(\tau_n(s_1), t_2)) \Delta_1 s_1, \tag{13}$$

$t_2 \in [a, b]$, or

$$\frac{\partial_2}{\Delta_2 t_2} \int_{t_1^0}^{t_1} u(\tau_0(s_1), t_2) \Delta_1 s_1 = p(t_2) \int_{t_1^0}^{t_1} u(\tau_0(s_1), t_2) \Delta_1 s_1$$

$$-p(t_2) \int_{t_1^0}^{t_1} u(\tau_0(s_1), t_2) \Delta_1 s_1 - u(t_1, t_2) + u(t_1^0, t_2)$$

$$- \sum_{j=0}^{n} \int_{t_1^0}^{t_1} b_j(s_1) u(\tau_j(s_1), t_2) \Delta_1 s_1$$

$$+ \int_{t_1^0}^{t_1} f(s_1, t_2, u(s_1, t_2), u(\tau_0(s_1), t_2), \ldots, u(\tau_n(s_1), t_2)) \Delta_1 s_1,$$
(14)

$t_2 \in [a, b]$. Hence and Lemma 1, we get (12).

2. Let $u \in \mathcal{C}_{rd}^1([t_1^0, \infty) \times [a, b])$ be a solution of (12). By Lemma 1, we obtain (14), whereupon (13). Hence, differentiating (13) with respect to t_1, we get (7). By (14) and Lemma 1, we have

$$\int_{t_1^0}^{t_1} u(\tau_0(s_1), a) \Delta_1 s_1 = \int_{t_1^0}^{t_1} u(\tau_0(s_1), b) \Delta_1 s_1, \quad t_1 \in [t_1^0, \infty).$$

We differentiate the last equation with respect to t_1 and we obtain (8). This completes the proof.

$\square$

Lemma 3. *Suppose* (G1)-(G5). *Then* $u \in \mathcal{C}_{rd}^1([t_1^0, \infty) \times [a, b])$ *is a solution of the BVP* (7), (8) *if and only if it satisfies the integral equation*

$$\int_{t_1^0}^{t_1} \int_{t_1^0}^{y_1} \int_{t_1^0}^{s_1} \int_a^{t_2} \int_a^{y_2} u(\tau_0(z_1), z_2) \Delta_2 z_2 \Delta_2 y_2 \Delta_1 z_1 \Delta_1 s_1 \Delta_1 y_1$$

$$= \int_{t_1^0}^{t_1} \int_{t_1^0}^{y_1} \int_a^{t_2} \int_a^{y_2} \int_a^b G(z_2, s_2) \Big(- p(s_2) \int_{t_1^0}^{z_1} u(\tau_0(s_1), s_2) \Delta_1 s_1$$

$$- u(z_1, s_2) + u(t_1^0, s_2)$$
(15)

$$- \sum_{j=0}^{n} \int_{t_1^0}^{z_1} b_j(s_1) u(\tau_j(s_1), s_2) \Delta_1 s_1$$

$$+ \int_{t_1^0}^{z_1} f(s_1, s_2, u(s_1, s_2), u(\tau_0(s_1), s_2), \ldots, u(\tau_n(s_1), s_2)) \Delta_1 s_1 \Big)$$

$$\Delta_2 s_2 \Delta_2 z_2 \Delta_2 y_2 \Delta_1 z_1 \Delta_1 y_1,$$

$t_1 \in [t_1^0, \infty)$, $t_2 \in [a, b]$.

Proof. 1. Let $u \in \mathcal{C}_{rd}^1([t_1^0, \infty) \times [a, b])$ is a solution of the BVP (7), (8). Then, using Lemma 2, we get (12) which we integrate twice in t_2 and then twice t_1 and we get (15).

2. Let $u \in \mathcal{C}_{rd}^1([t_1^0, \infty) \times [a, b])$ is a solution of (15). We differentiate (15) twice in t_1 and twice in t_2 and we get (12). Hence and Lemma 2, we obtain that $u \in \mathcal{C}_{rd}^1([t_1^0, \infty) \times [a, b])$ is a solution of the BVP (7), (8). This completes the proof. □

Theorem 18. *Suppose (G1)-(G5). Then the IBVP (7)-(9) has a solution $u \in \mathcal{C}_{rd}^1([t_1^0, \infty) \times [a, b])$.*

Proof. We will use the method of steps.

1. Let t_1^1 is the smallest value of $t_1 \in [t_1^0, \infty)$ such that
$$\tau_j(t) \in [t_1^{-1}, t_1^0], \quad t \in [t_1^0, t_1^1], \quad j \in \{0, 1, \ldots, n\}.$$

Let also, $E_1 = \mathcal{C}_{rd}^1([t_1^0, t_1^1] \times [a, b])$ be endowed with the norm
$$\|u\| = \max\left\{ \max_{(t_1, t_2) \in [t_1^0, t_1^1]} |u(t_1, t_2)|, \max_{(t_1, t_2) \in [t_1^0, t_1^1]} \left|\frac{\partial_1}{\Delta_1 t_1} u(t_1, t_2)\right|, \right.$$
$$\left. \max_{(t_1, t_2) \in [t_1^0, t_1^1]} \left|\frac{\partial_2}{\Delta_2 t_2} u(t_1, t_2)\right| \right\}.$$

Note that E_1 is a Banach space. Let A_1 be a positive constant such that
$$|\phi(t_1, t_2)| \leq A_1, \quad (t_1, t_2) \in [t_1^0, t_1^1] \times [a, b],$$
$$c(t_1, t_2) \leq A_1, \quad (t_1, t_2) \in [t_1^0, t_1^1] \times [a, b],$$
$$c_j(t_1, t_2) \leq A_1, \quad (t_1, t_2) \in [t_1^0, t_1^1] \times [a, b], \quad j \in \{0, 1, \ldots, n\},$$
$$G(t_2, s_2) \leq A_1, \quad t_2, s_2 \in [a, b],$$
$$|p(t_2)| \leq A_1, \quad t_2 \in [a, b],$$
$$|b_j(t_1)| \leq A_1, \quad t_1 \in [t_1^0, t_1^1], \quad j \in \{0, 1, \ldots, n\}.$$

Let $\tilde{\tilde{X}}_1$ be the set of all equi-continuous families in E_1, $\tilde{\tilde{X}}_1 = \tilde{\tilde{X}}_1 \bigcup \{\phi\}$, $\tilde{X}_1 = \overline{\tilde{\tilde{X}}_1}$, i.e., $\tilde{X}_1$ is the completion of $\tilde{\tilde{X}}_1$, and

$$X_1 = \{u \in \tilde{X}_1 : \|u\| \leq A_1\}.$$

Next,

$$|f(s_1, s_2, u(s_1, s_2), \phi(\tau_0(s_1), s_2), \ldots, \phi(\tau_n(s_1), s_2))|$$

$$\leq c(s_1, s_2)|u(s_1, s_2)|^l + \sum_{j=0}^{n} c_j(s_1, s_2)|\phi(\tau_j(s_1), s_2)|^{l_j}$$

$$\leq A_1^{1+l} + \sum_{j=0}^{n} A_1^{1+l_j}$$

$$= B_1$$

for any $(s_1, s_2) \in [t_1^0, t_1^1]$ and for any $u \in X_1$. Take $\epsilon_1 \in (0, 1)$ small enough so that

$$A_1 \geq \epsilon_1 \bigg(A_1 + A_1(t_1^1 - t_1^0)^3(b-a)^2 + 2A_1(t_1^1 - t_1^0)^2(b-a)^3$$

$$+ (A_1^3 + (n+1)A_1 + B_1)(t_1^1 - t_1^0)^3(b-a)^3 \bigg),$$

$$A_1 \geq \epsilon_1 \bigg(A_1 + A_1(t_1^1 - t_1^0)^2(b-a)^2 + 2A_1(t_1^1 - t_1^0)(b-a)^3$$

$$+ (A_1^3 + (n+1)A_1 + B_1)(t_1^1 - t_1^0)^2(b-a)^3 \bigg),$$

$$A_1 \geq \epsilon_1 \bigg(A_1 + A_1(t_1^1 - t_1^0)^3(b-a) + 2A_1(t_1^1 - t_1^0)^2(b-a)^2$$

$$+ (A_1^3 + (n+1)A_1 + B_1)(t_1^1 - t_1^0)^3(b-a)^2 \bigg).$$

Let $C_1 \geq \frac{1+\epsilon_1}{\epsilon_1} A_1$ and

$$Y_1 = \{u \in \tilde{X}_1 : \|u\| \leq (1+\epsilon_1)A_1\}.$$

Note that X_1 and Y_1 are compact subsets of E_1 and $X_1 \subset Y_1$. Also,
$$C_1 \geq (1+\epsilon_1)A_1$$
and
$$\frac{C_1}{1+\epsilon_1} + A_1 \leq C_1 \iff$$
$$A_1 \leq C_1 - \frac{C_1}{1+\epsilon_1}$$
$$\leq \frac{C_1(1+\epsilon_1 - 1)}{1+\epsilon_1}$$
$$= \frac{C_1 \epsilon_1}{1+\epsilon_1}.$$

For $u \in E_1$, define the operators
$$S_1 u(t_1, t_2) = (1+\epsilon_1) u(t_1, t_2),$$

$$T_1 u(t_1,t_2) = -\epsilon_1 \bigg(u(t_1,t_2)$$
$$+ \int_{t_1^0}^{t_1} \int_{t_1^0}^{y_1} \int_{t_1^0}^{s_1} \int_a^{t_2} \int_a^{y_2} u(\tau_0(z_1), z_2) \Delta_2 z_2 \Delta_2 y_2 \Delta_1 z_1 \Delta_1 s_1 \Delta_1 y_1$$
$$- \int_{t_1^0}^{t_1} \int_{t_1^0}^{y_1} \int_a^{t_2} \int_a^{y_2} \int_a^b G(z_2, s_2)\bigg(-p(s_2) \int_{t_1^0}^{z_1} u(\tau_0(s_1), s_2) \Delta_1 s_1$$
$$-u(z_1, s_2) + u(t_1^0, s_2)$$
$$- \sum_{j=0}^n \int_{t_1^0}^{z_1} b_j(s_1) u(\tau_j(s_1), s_2) \Delta_1 s_1$$
$$+ \int_{t_1^0}^{z_1} f(s_1, s_2, u(s_1, s_2), u(\tau_0(s_1), s_2), \ldots, u(\tau_n(s_1), s_2)) \Delta_1 s_1 \bigg)$$
$$\Delta_2 s_2 \Delta_2 z_2 \Delta_2 y_2 \Delta_1 z_1 \Delta_1 y_1 \bigg),$$

$(t_1, t_2) \in [t_1^0, t_1^1] \times [a, b].$

(a) Observe that, for $u_1, u_2 \in X_1$, we have

$$|S_1 u_1(t_1, t_2) - S_1 u_2(t_1, t_2)| = (1 + \epsilon_1)|u_1(t_1, t_2) - u_2(t_1, t_2)|,$$

$$\left|\frac{\partial_1}{\Delta_1 t_1} S_1 u_1(t_1, t_2) - \frac{\partial_1}{\Delta_1 t_1} S_1 u_2(t_1, t_2)\right| = (1 + \epsilon_1)$$
$$\times \left|\frac{\partial_1}{\Delta_1 t_1} u_1(t_1, t_2) - \frac{\partial_1}{\Delta_1 t_1} u_2(t_1, t_2)\right|,$$

$$\left|\frac{\partial_2}{\Delta_2 t_2} S_1 u_1(t_1, t_2) - \frac{\partial_2}{\Delta_2 t_2} S_1 u_2(t_1, t_2)\right| = (1 + \epsilon_1)$$
$$\times \left|\frac{\partial_2}{\Delta_2 t_2} u_1(t_1, t_2) - \frac{\partial_2}{\Delta_2 t_2} u_2(t_1, t_2)\right|,$$

$(t_1, t_2) \in [t_1^0, t_1^1] \times [a, b]$. Thus, $S_1 : E_1 \to E_1$ is a linear continuous operator which is expansive with a constant $(1 + \epsilon_1)$. Next, for $u \in X_1$, we get

$$\|S_1 u\| = (1 + \epsilon_1)\|u\|$$
$$\leq (1 + \epsilon_1) A_1,$$

i.e., $S_1(X_1) \subseteq Y_1$.

(b) Let $u \in X_1$ be arbitrarily chosen. We have

$$|T_1 u(t_1, t_2)| = \left| -\epsilon_1 \Bigg(u(t_1, t_2) \right.$$
$$+ \int_{t_1^0}^{t_1} \int_{t_1^0}^{y_1} \int_{t_1^0}^{s_1} \int_a^{t_2} \int_a^{y_2} \phi(\tau_0(z_1), z_2) \Delta_2 z_2 \Delta_2 y_2 \Delta_1 z_1 \Delta_1 s_1 \Delta_1 y_1$$
$$- \int_{t_1^0}^{t_1} \int_{t_1^0}^{y_1} \int_a^{t_2} \int_a^{y_2} \int_a^{b} G(z_2, s_2) \left(-p(s_2) \int_{t_1^0}^{z_1} \phi(\tau_0(s_1), s_2) \Delta_1 s_1 \right.$$
$$-u(z_1, s_2) + u(t_1^0, s_2)$$
$$- \sum_{j=0}^{n} \int_{t_1^0}^{z_1} b_j(s_1) \phi(\tau_j(s_1), s_2) \Delta_1 s_1$$
$$+ \int_{t_1^0}^{z_1} f(s_1, s_2, u(s_1, s_2), \phi(\tau_0(s_1), s_2), \ldots, \phi(\tau_n(s_1), s_2))$$
$$\left. \Delta_1 s_1 \right) \Delta_2 s_2 \Delta_2 z_2 \Delta_2 y_2 \Delta_1 z_1 \Delta_1 y_1 \Bigg) \Bigg|$$

$$\leq \epsilon_1 \bigg(|u(t_1, t_2)|$$

$$+ \bigg| \int_{t_1^0}^{t_1} \int_{t_1^0}^{y_1} \int_{t_1^0}^{s_1} \int_{a}^{t_2} \int_{a}^{y_2} \phi(\tau_0(z_1), z_2) \Delta_2 z_2 \Delta_2 y_2 \Delta_1 z_1 \Delta_1 s_1 \Delta_1 y_1$$

$$- \int_{t_1^0}^{t_1} \int_{t_1^0}^{y_1} \int_{a}^{t_2} \int_{a}^{y_2} \int_{a}^{b} G(z_2, s_2) \bigg(- p(s_2) \int_{t_1^0}^{z_1} \phi(\tau_0(s_1), s_2) \Delta_1 s_1$$

$$- u(z_1, s_2) + u(t_1^0, s_2)$$

$$- \sum_{j=0}^{n} \int_{t_1^0}^{z_1} b_j(s_1) \phi(\tau_j(s_1), s_2) \Delta_1 s_1$$

$$+ \int_{t_1^0}^{z_1} f(s_1, s_2, u(s_1, s_2), \phi(\tau_0(s_1), s_2), \ldots, \phi(\tau_n(s_1), s_2))$$

$$\Delta_1 s_1 \bigg) \Delta_2 s_2 \Delta_2 z_2 \Delta_2 y_2 \Delta_1 z_1 \Delta_1 y_1 \bigg| \bigg)$$

$$\leq \epsilon_1 \bigg(|u(t_1, t_2)|$$

$$+ \int_{t_1^0}^{t_1} \int_{t_1^0}^{y_1} \int_{t_1^0}^{s_1} \int_{a}^{t_2} \int_{u}^{y_2} |\phi(\tau_0(z_1), z_2)| \Delta_2 z_2 \Delta_2 y_2 \Delta_1 z_1 \Delta_1 s_1 \Delta_1 y_1$$

$$+ \int_{t_1^0}^{t_1} \int_{t_1^0}^{y_1} \int_{a}^{t_2} \int_{a}^{y_2} \int_{a}^{b} G(z_2, s_2) \bigg(|p(s_2)| \int_{t_1^0}^{z_1} |\phi(\tau_0(s_1), s_2)| \Delta_1 s_1$$

$$+ |u(z_1, s_2)| + |u(t_1^0, s_2)|$$

$$+ \sum_{j=0}^{n} \int_{t_1^0}^{z_1} |b_j(s_1)| |\phi(\tau_j(s_1), s_2)| \Delta_1 s_1$$

$$+ \int_{t_1^0}^{z_1} |f(s_1, s_2, u(s_1, s_2), \phi(\tau_0(s_1), s_2), \ldots, \phi(\tau_n(s_1), s_2))| \Delta_1 s_1 \bigg)$$

$$\Delta_2 s_2 \Delta_2 z_2 \Delta_2 y_2 \Delta_1 z_1 \Delta_1 y_1 \bigg)$$

$$\begin{aligned}
&\leq \epsilon_1\Big(A_1 + A_1(t_1^1 - t_1^0)^3(b-a)^2 + A_1^3(t_1^1 - t_1^0)^3(b-a)^3 \\
&\quad + 2A_1(t_1^1 - t_1^0)^2(b-a)^3 + (n+1)A_1(t_1^1 - t_1^0)^3(b-a)^3 \\
&\quad + B_1(t_1^1 - t_1^0)^3(b-a)^3\Big) \\
&= \epsilon_1\Big(A_1 + A_1(t_1^1 - t_1^0)^3(b-a)^2 + 2A_1(t_1^1 - t_1^0)^2(b-a)^3 \\
&\quad + (A_1^3 + (n+1)A_1 + B_1)(t_1^1 - t_1^0)^3(b-a)^3\Big) \\
&\leq A_1,
\end{aligned}$$

$(t_1, t_2) \in [t_1^0, t_1^1] \times [a, b]$. Next,

$$\left|\frac{\partial_1}{\Delta_1 t_1} T_1 u(t_1, t_2)\right| = \Bigg| - \epsilon_1\Bigg(\frac{\partial_1}{\Delta_1 t_1} u(t_1, t_2)$$

$$+ \int_{t_1^0}^{t_1} \int_{t_1^0}^{s_1} \int_a^{t_2} \int_a^{y_2} \phi(\tau_0(z_1), z_2) \Delta_2 z_2 \Delta_2 y_2 \Delta_1 z_1 \Delta_1 s_1$$

$$- \int_{t_1^0}^{t_1} \int_a^{t_2} \int_a^{y_2} \int_a^b G(z_2, s_2)\Bigg(-p(s_2)\int_{t_1^0}^{z_1} \phi(\tau_0(s_1), s_2)\Delta_1 s_1$$

$$- u(z_1, s_2) + u(t_1^0, s_2)$$

$$- \sum_{j=0}^n \int_{t_1^0}^{z_1} b_j(s_1)\phi(\tau_j(s_1), s_2)\Delta_1 s_1$$

$$+ \int_{t_1^0}^{z_1} f(s_1, s_2, u(s_1, s_2), \phi(\tau_0(s_1), s_2), \ldots, \phi(\tau_n(s_1), s_2))$$

$$\Delta_1 s_1\Bigg) \Delta_2 s_2 \Delta_2 z_2 \Delta_2 y_2 \Delta_1 z_1\Bigg)\Bigg|$$

$$\leq \epsilon_1 \left(\left| \frac{\partial_1}{\Delta_1 t_1} u(t_1, t_2) \right| \right.$$

$$+ \left| \int_{t_1^0}^{t_1} \int_{t_1^0}^{s_1} \int_a^{t_2} \int_a^{y_2} \phi(\tau_0(z_1), z_2) \Delta_2 z_2 \Delta_2 y_2 \Delta_1 z_1 \Delta_1 s_1 \right|$$

$$+ \left| \int_{t_1^0}^{t_1} \int_a^{t_2} \int_a^{y_2} \int_a^{b} G(z_2, s_2) \left(-p(s_2) \int_{t_1^0}^{z_1} \phi(\tau_0(s_1), s_2) \Delta_1 s_1 \right. \right.$$

$$-u(z_1, s_2) + u(t_1^0, s_2)$$

$$- \sum_{j=0}^{n} \int_{t_1^0}^{z_1} b_j(s_1) \phi(\tau_j(s_1), s_2) \Delta_1 s_1$$

$$+ \int_{t_1^0}^{z_1} f\left(s_1, s_2, u(s_1, s_2), \phi(\tau_0(s_1), s_2), \ldots, \phi(\tau_n(s_1), s_2)\right)$$

$$\left. \left. \Delta_1 s_1 \right) \Delta_2 s_2 \Delta_2 z_2 \Delta_2 y_2 \Delta_1 z_1 \right) \right|$$

$$\leq \epsilon_1 \left(\left| \frac{\partial_1}{\Delta_1 t_1} u(t_1, t_2) \right| \right.$$

$$+ \int_{t_1^0}^{t_1} \int_{t_1^0}^{s_1} \int_a^{t_2} \int_a^{y_2} |\phi(\tau_0(z_1), z_2)| \Delta_2 z_2 \Delta_2 y_2 \Delta_1 z_1 \Delta_1 s_1 \right|$$

$$+ \int_{t_1^0}^{t_1} \int_a^{t_2} \int_a^{y_2} \int_a^{b} G(z_2, s_2) \left(|p(s_2)| \int_{t_1^0}^{z_1} |\phi(\tau_0(s_1), s_2)| \Delta_1 s_1 \right.$$

$$+ |u(z_1, s_2)| + |u(t_1^0, s_2)|$$

$$+ \sum_{j=0}^{n} \int_{t_1^0}^{z_1} |b_j(s_1)| |\phi(\tau_j(s_1), s_2)| \Delta_1 s_1$$

$$+ \int_{t_1^0}^{z_1} |f\left(s_1, s_2, u(s_1, s_2), \phi(\tau_0(s_1), s_2), \ldots, \phi(\tau_n(s_1), s_2)\right)|$$

$$\left. \Delta_1 s_1 \right) \Delta_2 s_2 \Delta_2 z_2 \Delta_2 y_2 \Delta_1 z_1 \right)$$

$$\leq \epsilon_1 \Big(A_1 + A_1(t_1^1 - t_1^0)^2(b-a)^2 + A_1^3(t_1^1 - t_1^0)^2(b-a)^3$$

$$+ 2A_1(t_1^1 - t_1^0)(b-a)^3 + (n+1)A_1(t_1^1 - t_1^0)^2(b-a)^3$$

$$+ B_1(t_1^1 - t_1^0)^2(b-a)^3 \Big)$$

$$= \epsilon_1 \Big(A_1 + A_1(t_1^1 - t_1^0)^2(b-a)^2 + 2A_1(t_1^1 - t_1^0)(b-a)^3$$

$$+ (A_1^3 + (n+1)A_1 + B_1)(t_1^1 - t_1^0)^2(b-a)^3 \Big)$$

$$\leq A_1,$$

$(t_1, t_2) \in [t_1^0, t_1^1] \times [a, b]$, and

$$\left| \frac{\partial_2}{\Delta_2 t_2} T_1 u(t_1, t_2) \right| = \left| -\epsilon_1 \Big(\frac{\partial_2}{\Delta_2 t_2} u(t_1, t_2) \right.$$

$$+ \int_{t_1^0}^{t_1} \int_{t_1^0}^{y_1} \int_{t_1^0}^{s_1} \int_a^{t_2} \phi(\tau_0(z_1), z_2) \Delta_2 z_2 \Delta_1 z_1 \Delta_1 s_1 \Delta_1 y_1$$

$$- \int_{t_1^0}^{t_1} \int_{t_1^0}^{y_1} \int_a^{t_2} \int_a^b G(z_2, s_2) \Big(-p(s_2) \int_{t_1^0}^{z_1} \phi(\tau_0(s_1), s_2) \Delta_1 s_1$$

$$- u(z_1, s_2) + u(t_1^0, s_2)$$

$$- \sum_{j=0}^n \int_{t_1^0}^{z_1} b_j(s_1) \phi(\tau_j(s_1), s_2) \Delta_1 s_1$$

$$+ \int_{t_1^0}^{z_1} f(s_1, s_2, u(s_1, s_2), \phi(\tau_0(s_1), s_2), \ldots, \phi(\tau_n(s_1), s_2)) \Delta_1 s_1 \Big)$$

$$\Delta_2 s_2 \Delta_2 y_2 \Delta_1 z_1 \Delta_1 y_1 \Big) \Big|$$

$$\leq \epsilon_1 \Bigg(\bigg| \frac{\partial_2}{\Delta_2 t_2} u(t_1, t_2) \bigg|$$

$$+ \bigg| \int_{t_1^0}^{t_1} \int_{t_1^0}^{y_1} \int_{t_1^0}^{s_1} \int_a^{t_2} \phi(\tau_0(z_1), z_2) \Delta_2 z_2 \Delta_1 z_1 \Delta_1 s_1 \Delta_1 y_1$$

$$- \int_{t_1^0}^{t_1} \int_{t_1^0}^{y_1} \int_a^{t_2} \int_a^b G(z_2, s_2) \bigg(-p(s_2) \int_{t_1^0}^{z_1} \phi(\tau_0(s_1), s_2) \Delta_1 s_1$$

$$-u(z_1, s_2) + u(t_1^0, s_2)$$

$$- \sum_{j=0}^n \int_{t_1^0}^{z_1} b_j(s_1) \phi(\tau_j(s_1), s_2) \Delta_1 s_1$$

$$+ \int_{t_1^0}^{z_1} f\left(s_1, s_2, u(s_1, s_2), \phi(\tau_0(s_1), s_2), \ldots, \phi(\tau_n(s_1), s_2)\right)$$

$$\Delta_1 s_1 \bigg) \Delta_2 s_2 \Delta_2 z_2 \Delta_1 z_1 \Delta_1 y_1 \bigg) \bigg|$$

$$\leq \epsilon_1 \Bigg(\bigg| \frac{\partial_2}{\Delta_2 t_2} u(t_1, t_2) \bigg|$$

$$+ \int_{t_1^0}^{t_1} \int_{t_1^0}^{y_1} \int_{t_1^0}^{s_1} \int_a^{t_2} |\phi(\tau_0(z_1), z_2)| \Delta_2 z_2 \Delta_1 z_1 \Delta_1 s_1 \Delta_1 y_1$$

$$+ \int_{t_1^0}^{t_1} \int_{t_1^0}^{y_1} \int_a^{t_2} \int_a^b G(z_2, s_2) \bigg(|p(s_2)| \int_{t_1^0}^{z_1} |\phi(\tau_0(s_1), s_2)| \Delta_1 s_1$$

$$+ |u(z_1, s_2)| + |u(t_1^0, s_2)|$$

$$+ \sum_{j=0}^n \int_{t_1^0}^{z_1} |b_j(s_1)| |\phi(\tau_j(s_1), s_2)| \Delta_1 s_1$$

$$+ \int_{t_1^0}^{z_1} |f\left(s_1, s_2, u(s_1, s_2), \phi(\tau_0(s_1), s_2), \ldots, \phi(\tau_n(s_1), s_2)\right)|$$

$$\Delta_1 s_1 \bigg) \Delta_2 s_2 \Delta_2 z_2 \Delta_1 z_1 \Delta_1 y_1 \bigg)$$

$$\leq \epsilon_1 \bigg(A_1 + A_1(t_1^1 - t_1^0)^3 (b-a) + A_1^3 (t_1^1 - t_1^0)^3 (b-a)^2$$

$$+ 2A_1 (t_1^1 - t_1^0)^2 (b-a)^2 + (n+1) A_1 (t_1^1 - t_1^0)^3 (b-a)^2$$

$$+ B_1 (t_1^1 - t_1^0)^3 (b-a)^2 \bigg)$$

$$= \epsilon_1 \bigg(A_1 + A_1 (t_1^1 - t_1^0)^3 (b-a) + 2A_1 (t_1^1 - t_1^0)^2 (b-a)^2$$

$$+(A_1^3 + (n+1)A_1 + B_1)(t_1^1 - t_1^0)^3(b-a)^2\Big)$$

$$\leq A_1,$$

$(t_1, t_2) \in [t_1^0, t_1^1] \times [a, b]$. Therefore $T_1(X_1) \subset X_1$.

Let $z_1 \in T_1(X_1)$ and $y \in Y_1$. Then $\|z\| \leq A_1$ and

$$\|S_1^{-1} y - z\| \leq \|S_1^{-1} y\| + \|z\|$$

$$\leq \frac{\|y\|}{1 + \epsilon_1} + A_1$$

$$\leq \frac{C_1}{1 + \epsilon_1} + A_1$$

$$\leq C_1,$$

i.e., $S_1^{-1} y - z \in Y_1$ and

$$\{S_1^{-1} y - z : y \in Y_1, \quad z \in T_1(X_1)\} \subset Y_1.$$

Hence and Theorem 17, we conclude that the operator $S_1 + T_1$ has a fixed point $u_1 \in X_1$. From here, $u_1 \in \mathcal{C}_{rd}^1([t_1^0, t_1^1] \times [a, b])$ is a solution of the IBVP

$$u_{t_1}^{\Delta_1}(t_1, t_2) + b_0(t_1) u_{t_2}^{\Delta_2}(\tau_0(t_1), t_2) + \sum_{j=1}^{n} b_j(t_1) u(\tau_j(t_1), t_2)$$

$$= f(t_1, t_2, u(t_1, t_2), u(\tau_0(t_1), t_2), u(\tau_1(t_1), t_2), \ldots, u(\tau_n(t_1), t_2)),$$

$$t_1 \in [t_1^0, t_1^1], \quad t_2 \in [a, b],$$
$$u(t_1, a) = u(t_1, b), \quad t_1 \in [t_1^0, t_1^1],$$
$$u(t_1, t_2) = \phi(t_1, t_2), \quad t_1 \in [t_1^{-1}, t_1^0], \quad t_2 \in [a, b].$$

2. Let t_1^2 is the smallest value of $t_1 \in [t_1^1, \infty)$ such that

$$\tau_j(t) \in [t_1^0, t_1^1], \quad t \in [t_1^1, t_1^2], \quad j \in \{0, 1, \ldots, n\}.$$

Let also, $E_2 = C_{rd}^1([t_1^1, t_1^2] \times [a, b])$ be endowed with the norm

$$\|u\| = \max\left\{\max_{(t_1,t_2)\in[t_1^1,t_1^2]} |u(t_1, t_2)|, \max_{(t_1,t_2)\in[t_1^1,t_1^2]} \left|\frac{\partial_1}{\Delta_1 t_1} u(t_1, t_2)\right|,\right.$$

$$\left.\max_{(t_1,t_2)\in[t_1^1,t_1^2]} \left|\frac{\partial_2}{\Delta_2 t_2} u(t_1, t_2)\right|\right\}.$$

Note that E_2 is a Banach space. Let $A_2 \geq A_1$ be a positive constant such that

$$|u_1(t_1, t_2)| \leq A_2, \quad (t_1, t_2) \in [t_1^0, t_1^1] \times [a, b],$$

$$c(t_1, t_2) \leq A_2, \quad (t_1, t_2) \in [t_1^1, t_1^2] \times [a, b],$$

$$c_j(t_1, t_2) \leq A_2, \quad (t_1, t_2) \in [t_1^1, t_1^2] \times [a, b], \quad j \in \{0, 1, \ldots, n\},$$

$$G(t_2, s_2) \leq A_2, \quad t_2, s_2 \in [a, b],$$

$$|p(t_2)| \leq A_2, \quad t_2 \in [a, b],$$

$$|b_j(t_1)| \leq A_2, \quad t_1 \in [t_1^1, t_1^2], \quad j \in \{0, 1, \ldots, n\}.$$

Let $\widetilde{\widetilde{X}}_2$ be the set of all equi-continuous families in E_2, $\widetilde{\widetilde{X}}_2 = \widetilde{\widetilde{X}}_2 \bigcup \{\phi\}$, $\widetilde{X}_2 = \overline{\widetilde{\widetilde{X}}_2}$, i.e., $\widetilde{X}_2$ is the completion of $\widetilde{\widetilde{X}}_2$, and

$$X_2 = \{x \in \widetilde{X}_2 : \|u\| \leq A_2\}.$$

Also,

$$|f(s_1, s_2, u(s_1, s_2), u_1(\tau_0(s_1), s_2), \ldots, u_1(\tau_n(s_1), s_2))|$$

$$\leq c(s_1, s_2)|u(s_1, s_2)|^l + \sum_{j=0}^{n} c_j(s_1, s_2)|u_1(\tau_j(s_1), s_2)|^{l_j}$$

$$\leq A_2^{1+l} + \sum_{j=0}^{n} A_2^{1+l_j}$$

$$= B_2$$

for any $(s_1, s_2) \in [t_1^1, t_1^2]$ and for any $u \in X_2$. Take $\epsilon_2 \in (0, 1)$ so that

$$A_2 \geq \epsilon_2 \bigg(A_2 + A_2(t_1^2 - t_1^1)^3(b-a)^2 + 2A_2(t_1^2 - t_1^1)^2(b-a)^3$$

$$+ (A_2^3 + (n+1)A_2 + B_2)(t_1^2 - t_1^1)^3(b-a)^3 \bigg),$$

$$A_2 \geq \epsilon_2 \bigg(A_2 + A_2(t_1^2 - t_1^1)^2(b-a)^2 + 2A_2(t_1^2 - t_1^1)(b-a)^3$$

$$+ (A_2^3 + (n+1)A_2 + B_2)(t_1^2 - t_1^1)^2(b-a)^3 \bigg),$$

$$A_2 \geq \epsilon_2 \bigg(A_2 + A_2(t_1^2 - t_1^1)^3(b-a) + 2A_2(t_1^2 - t_1^1)^2(b-a)^2$$

$$+ (A_2^3 + (n+1)A_2 + B_2)(t_1^2 - t_1^1)^3(b-a)^2 \bigg).$$

Let $C_2 \geq \frac{1+\epsilon_2}{\epsilon_2} A_2$ and

$$\frac{C_2}{1+\epsilon_2} + A_2 \leq C_2 \iff$$

$$A_2 \leq C_2 - \frac{C_2}{1+\epsilon_2}$$

$$= \frac{\epsilon_2 C_2}{1+\epsilon_2}.$$

$$Y_2 = \{u \in \widetilde{X}_{\text{''}} : \|u\| \leq (1+\epsilon_2) A_2\}.$$

Note that $X_2 \subset Y_2$ and X_2 and Y_2 are compact subsets of E_2. For $u \in E_2$, define the operators

$$S_2 u(t_1, t_2) = (1+\epsilon_2) u(t_1, t_2),$$

$$T_2 u(t_1, t_2) = -\epsilon_2 \bigg(u(t_1, t_2)$$

$$+ \int_{t_1^1}^{t_1} \int_{t_1^1}^{y_1} \int_{t_1^1}^{s_1} \int_a^{t_2} \int_a^{y_2} u(\tau_0(z_1), z_2) \Delta_2 z_2 \Delta_2 y_2 \Delta_1 z_1 \Delta_1 s_1 \Delta_1 y_1$$

$$- \int_{t_1^1}^{t_1} \int_{t_1^1}^{y_1} \int_a^{t_2} \int_a^{y_2} \int_a^b G(z_2, s_2) \bigg(-p(s_2) \int_{t_1^1}^{z_1} u(\tau_0(s_1), s_2) \Delta_1 s_1$$

$$- u(z_1, s_2) + u(t_1^1, s_2)$$

$$- \sum_{j=0}^n \int_{t_1^1}^{z_1} b_j(s_1) u(\tau_j(s_1), s_2) \Delta_1 s_1$$

$$+ \int_{t_1^1}^{z_1} f(s_1, s_2, u(s_1, s_2), u(\tau_0(s_1), s_2), \ldots, u(\tau_n(s_1), s_2))$$

$$\Delta_1 s_1 \bigg) \Delta_2 s_2 \Delta_2 z_2 \Delta_2 y_2 \Delta_1 z_1 \Delta_1 y_1 \bigg),$$

$(t_1, t_2) \in [t_1^1, t_1^2] \times [a, b]$.

(a) Observe that, for $u_1, u_2 \in X_2$, we have

$$|S_2 u_1(t_1, t_2) - S_2 u_2(t_1, t_2)| = (1 + \epsilon_2) |u_1(t_1, t_2) - u_2(t_1, t_2)|,$$

$$\left| \frac{\partial_1}{\Delta_1 t_1} S_2 u_1(t_1, t_2) - \frac{\partial_1}{\Delta_1 t_1} S_2 u_2(t_1, t_2) \right| = (1 + \epsilon_2)$$
$$\times \left| \frac{\partial_1}{\Delta_1 t_1} u_1(t_1, t_2) - \frac{\partial_1}{\Delta_1 t_1} u_2(t_1, t_2) \right|,$$

$$\left| \frac{\partial_2}{\Delta_2 t_2} S_2 u_1(t_1, t_2) - \frac{\partial_2}{\Delta_2 t_2} S_2 u_2(t_1, t_2) \right| = (1 + \epsilon_2)$$
$$\times \left| \frac{\partial_2}{\Delta_2 t_2} u_1(t_1, t_2) - \frac{\partial_2}{\Delta_2 t_2} u_2(t_1, t_2) \right|,$$

$(t_1, t_2) \in [t_1^1, t_1^2] \times [a, b]$. Thus, $S_2 : X_2 \to E_2$ is a linear continuous operator which is expansive with a constant $(1 + \epsilon_2)$. Next, for

$u \in X_2$, we get

$$\|S_2 u\| = (1+\epsilon_2)\|u\|$$
$$\leq (1+\epsilon_2)A_2,$$

i.e., $S_2(X_2) \subseteq Y_2$.

(b) Let $u \in X_2$ be arbitrarily chosen. We have

$$|T_2 u(t_1, t_2)| = \bigg| -\epsilon_2 \bigg(u(t_1, t_2)$$

$$+ \int_{t_1^1}^{t_1} \int_{t_1^1}^{y_1} \int_{t_1^1}^{s_1} \int_{a}^{t_2} \int_{a}^{y_2} u_1(\tau_0(z_1), z_2) \Delta_2 z_2 \Delta_2 y_2 \Delta_1 z_1 \Delta_1 s_1 \Delta_1 y_1$$

$$- \int_{t_1^1}^{t_1} \int_{t_1^1}^{y_1} \int_{a}^{t_2} \int_{a}^{y_2} \int_{a}^{b} G(z_2, s_2)\bigg(-p(s_2) \int_{t_1^0}^{z_1} u_1(\tau_0(s_1), s_2)\Delta_1 s_1$$

$$-u(z_1, s_2) + u(t_1^1, s_2)$$

$$- \sum_{j=0}^{n} \int_{t_1^1}^{z_1} b_j(s_1) u_1(\tau_j(s_1), s_2) \Delta_1 s_1$$

$$+ \int_{t_1^1}^{z_1} f\left(s_1, s_2, u(s_1, s_2), u_1(\tau_0(s_1), s_2), \ldots, u_1(\tau_n(s_1), s_2)\right) \Delta_1 s_1 \bigg)$$

$$\Delta_2 s_2 \Delta_2 z_2 \Delta_2 y_2 \Delta_1 z_1 \Delta_1 y_1 \bigg) \bigg|$$

$$\leq \epsilon_2 \bigg(|u(t_1, t_2)|$$

$$+ \bigg| \int_{t_1^1}^{t_1} \int_{t_1^1}^{y_1} \int_{t_1^1}^{s_1} \int_{a}^{t_2} \int_{a}^{y_2} u_1(\tau_0(z_1), z_2) \Delta_2 z_2 \Delta_2 y_2 \Delta_1 z_1 \Delta_1 s_1 \Delta_1 y_1$$

$$- \int_{t_1^1}^{t_1} \int_{t_1^1}^{y_1} \int_{a}^{t_2} \int_{a}^{y_2} \int_{a}^{b} G(z_2, s_2)\bigg(-p(s_2) \int_{t_1^1}^{z_1} u_1(\tau_0(s_1), s_2) \Delta_1 s_1$$

$$-u(z_1, s_2) + u(t_1^1, s_2)$$

$$- \sum_{j=0}^{n} \int_{t_1^1}^{z_1} b_j(s_1) u_1(\tau_j(s_1), s_2) \Delta_1 s_1$$

$$+ \int_{t_1^1}^{z_1} f\left(s_1, s_2, u(s_1, s_2), u_1(\tau_0(s_1), s_2), \ldots, u_1(\tau_n(s_1), s_2)\right)$$

$$\Delta_1 s_1 \bigg) \Delta_2 s_2 \Delta_2 z_2 \Delta_2 y_2 \Delta_1 z_1 \Delta_1 y_1 \bigg| \bigg)$$

$$\leq \epsilon_2 \Big(|u(t_1,t_2)|$$

$$+ \int_{t_1^1}^{t_1} \int_{t_1^1}^{y_1} \int_{t_1^1}^{s_1} \int_{a}^{t_2} \int_{a}^{y_2} |u_1(\tau_0(z_1),z_2)| \Delta_2 z_2 \Delta_2 y_2 \Delta_1 z_1 \Delta_1 s_1 \Delta_1 y_1$$

$$+ \int_{t_1^1}^{t_1} \int_{t_1^1}^{y_1} \int_{a}^{t_2} \int_{a}^{y_2} \int_{a}^{b} G(z_2,s_2) \Big(|p(s_2)| \int_{t_1^1}^{z_1} |u_1(\tau_0(s_1),s_2)| \Delta_1 s_1$$

$$+ |u(z_1,s_2)| + |u(t_1^1,s_2)|$$

$$+ \sum_{j=0}^{n} \int_{t_1^1}^{z_1} |b_j(s_1)| |u_1(\tau_j(s_1),s_2)| \Delta_1 s_1$$

$$+ \int_{t_1^1}^{z_1} |f(s_1,s_2,u(s_1,s_2),u_1(\tau_0(s_1),s_2),\ldots,u_1(\tau_n(s_1),s_2))| \Delta_1 s_1 \Big)$$

$$\Delta_2 s_2 \Delta_2 z_2 \Delta_2 y_2 \Delta_1 z_1 \Delta_1 y_1 \Big)$$

$$\leq \epsilon_2 \Big(A_2 + A_2(t_1^2 - t_1^1)^3(b-a)^2 + A_2^3(t_1^2 - t_1^1)^3(b-a)^3$$

$$+ 2A_2(t_1^2 - t_1^1)^2(b-a)^3 + (n+1)A_2(t_1^2 - t_1^1)^3(b-a)^3$$

$$+ B_2(t_1^2 - t_1^1)^3(b-a)^3 \Big)$$

$$= \epsilon_2 \Big(A_2 + A_2(t_1^2 - t_1^1)^3(b-a)^2 + 2A_2(t_1^2 - t_1^1)^2(b-a)^3$$

$$+ (A_2^3 + (n+1)A_2 + B_2)(t_1^2 - t_1^1)^3(b-a)^3 \Big)$$

$$\leq A_2,$$

$(t_1, t_2) \in [t_1^1, t_1^2] \times [a, b]$. Next,

$$\left|\frac{\partial_1}{\Delta_1 t_1} T_2 u(t_1, t_2)\right| = \left| -\epsilon_2 \left(\frac{\partial_1}{\Delta_1 t_1} u(t_1, t_2) \right.\right.$$

$$+ \int_{t_1^1}^{t_1} \int_{t_1^1}^{s_1} \int_a^{t_2} \int_a^{y_2} u_1(\tau_0(z_1), z_2) \Delta_2 z_2 \Delta_2 y_2 \Delta_1 z_1 \Delta_1 s_1$$

$$- \int_{t_1^1}^{t_1} \int_a^{t_2} \int_a^{y_2} \int_a^b G(z_2, s_2) \left(-p(s_2) \int_{t_1^1}^{z_1} u_1(\tau_0(s_1), s_2) \Delta_1 s_1 \right.$$

$$- u(z_1, s_2) + u(t_1^1, s_2)$$

$$- \sum_{j=0}^n \int_{t_1^1}^{z_1} b_j(s_1) u_1(\tau_j(s_1), s_2) \Delta_1 s_1$$

$$+ \int_{t_1^1}^{z_1} f(s_1, s_2, u(s_1, s_2), u_1(\tau_0(s_1), s_2), \ldots, u_1(\tau_n(s_1), s_2)) \Delta_1 s_1 \bigg)$$

$$\Delta_2 s_2 \Delta_2 z_2 \Delta_2 y_2 \Delta_1 z_1 \bigg) \bigg|$$

$$\leq \epsilon_2 \left(\left| \frac{\partial_1}{\Delta_1 t_1} u(t_1, t_2) \right| \right.$$

$$+ \left| \int_{t_1^1}^{t_1} \int_{t_1^1}^{s_1} \int_a^{t_2} \int_a^{y_2} u_1(\tau_0(z_1), z_2) \Delta_2 z_2 \Delta_2 y_2 \Delta_1 z_1 \Delta_1 s_1 \right|$$

$$+ \left| \int_{t_1^1}^{t_1} \int_a^{t_2} \int_a^{y_2} \int_a^b G(z_2, s_2) \left(-p(s_2) \int_{t_1^1}^{z_1} u_1(\tau_0(s_1), s_2) \Delta_1 s_1 \right.\right.$$

$$- u(z_1, s_2) + u(t_1^1, s_2)$$

$$- \sum_{j=0}^n \int_{t_1^1}^{z_1} b_j(s_1) u_1(\tau_j(s_1), s_2) \Delta_1 s_1$$

$$+ \int_{t_1^1}^{z_1} f(s_1, s_2, u(s_1, s_2), u_1(\tau_0(s_1), s_2), \ldots, u_1(\tau_n(s_1), s_2)) \Delta_1 s_1 \bigg)$$

$$\Delta_2 s_2 \Delta_2 z_2 \Delta_2 y_2 \Delta_1 z_1 \bigg) \bigg|$$

$$\leq \epsilon_2 \left(\left| \frac{\partial_1}{\Delta_1 t_1} u(t_1, t_2) \right| \right.$$

$$+ \int_{t_1^1}^{t_1} \int_{t_1^1}^{s_1} \int_a^{t_2} \int_a^{y_2} |u_1(\tau_0(z_1), z_2)| \Delta_2 z_2 \Delta_2 y_2 \Delta_1 z_1 \Delta_1 s_1 \bigg|$$

$$+ \int_{t_1^1}^{t_1} \int_a^{t_2} \int_a^{y_2} \int_a^b G(z_2, s_2) \left(|p(s_2)| \int_{t_1^1}^{z_1} |u_1(\tau_0(s_1), s_2)| \Delta_1 s_1 \right.$$

$$+ |u(z_1, s_2)| + |u(t_1^1, s_2)|$$

$$+ \sum_{j=0}^n \int_{t_1^1}^{z_1} |b_j(s_1)| |u_1(\tau_j(s_1), s_2)| \Delta_1 s_1$$

$$+ \int_{t_1^1}^{z_1} |f(s_1, s_2, u(s_1, s_2), u_1(\tau_0(s_1), s_2), \ldots, u_1(\tau_n(s_1), s_2))| \Delta_1 s_1 \right)$$

$$\Delta_2 s_2 \Delta_2 z_2 \Delta_2 y_2 \Delta_1 z_1 \bigg)$$

$$\leq \epsilon_2 \bigg(A_2 + A_2 (t_1^2 - t_1^1)^2 (b-a)^2 + A_2^3 (t_1^2 - t_1^1)^2 (b-a)^3$$

$$+ 2A_2 (t_1^2 - t_1^1)(b-a)^3 + (n+1) A_2 (t_1^2 - t_1^1)^2 (b-a)^3$$

$$+ B_2 (t_1^2 - t_1^1)^2 (b-a)^3 \bigg)$$

$$= \epsilon_2 \bigg(A_2 + A_2 (t_1^2 - t_1^1)^2 (b-a)^2 + 2A_2 (t_1^2 - t_1^1)(b-a)^3$$

$$+ (A_2^3 + (n+1) A_2 + B_2)(t_1^2 - t_1^1)^2 (b-a)^3 \bigg)$$

$$\leq A_2,$$

$(t_1, t_2) \in [t_1^1, t_1^2] \times [a, b]$, and

$$\left| \frac{\partial_2}{\Delta_2 t_2} T_1 u(t_1, t_2) \right| = \left| - \epsilon_2 \left(\frac{\partial_2}{\Delta_2 t_2} u(t_1, t_2) \right. \right.$$

$$+ \int_{t_1^1}^{t_1} \int_{t_1^1}^{y_1} \int_{t_1^1}^{s_1} \int_a^{t_2} u_1(\tau_0(z_1), z_2) \Delta_2 z_2 \Delta_1 z_1 \Delta_1 s_1 \Delta_1 y_1$$

$$- \int_{t_1^1}^{t_1} \int_{t_1^1}^{y_1} \int_a^{t_2} \int_a^b G(z_2, s_2) \bigg(-p(s_2) \int_{t_1^1}^{z_1} u_1(\tau_0(s_1), s_2) \Delta_1 s_1$$

$$- u(z_1, s_2) + u(t_1^1, s_2)$$

$$-\sum_{j=0}^{n}\int_{t_1^1}^{z_1}b_j(s_1)u_1(\tau_j(s_1),s_2)\Delta_1 s_1$$

$$+\int_{t_1^1}^{z_1}f(s_1,s_2,u(s_1,s_2),u_1(\tau_0(s_1),s_2),\ldots,u_1(\tau_n(s_1),s_2))\,\Delta_1 s_1\Bigg)$$

$$\Delta_2 s_2 \Delta_2 y_2 \Delta_1 z_1 \Delta_1 y_1\Bigg)\Bigg|$$

$$\leq \epsilon_2\Bigg(\bigg|\frac{\partial_2}{\Delta_2 t_2}u(t_1,t_2)\bigg|$$

$$+\bigg|\int_{t_1^1}^{t_1}\int_{t_1^1}^{y_1}\int_{t_1^1}^{s_1}\int_{a}^{t_2}u_1(\tau_0(z_1),z_2)\Delta_2 z_2 \Delta_1 z_1 \Delta_1 s_1 \Delta_1 y_1$$

$$-\int_{t_1^1}^{t_1}\int_{t_1^1}^{y_1}\int_{a}^{t_2}\int_{a}^{b}G(z_2,s_2)\bigg(-p(s_2)\int_{t_1^1}^{z_1}u(\tau_0(s_1),s_2)\Delta_1 s_1$$

$$-u(z_1,s_2)+u(t_1^1,s_2)$$

$$-\sum_{j=0}^{n}\int_{t_1^1}^{z_1}b_j(s_1)u_1(\tau_j(s_1),s_2)\Delta_1 s_1$$

$$+\int_{t_1^1}^{z_1}f(s_1,s_2,u(s_1,s_2),u_1(\tau_0(s_1),s_2),\ldots,u_1(\tau_n(s_1),s_2))\,\Delta_1 s_1\Bigg)$$

$$\Delta_2 s_2 \Delta_2 z_2 \Delta_1 z_1 \Delta_1 y_1\Bigg)\Bigg|$$

$$\leq \epsilon_2\Bigg(\bigg|\frac{\partial_2}{\Delta_2 t_2}u(t_1,t_2)\bigg|$$

$$+\int_{t_1^1}^{t_1}\int_{t_1^1}^{y_1}\int_{t_1^1}^{s_1}\int_{a}^{t_2}|u_1(\tau_0(z_1),z_2)|\Delta_2 z_2 \Delta_1 z_1 \Delta_1 s_1 \Delta_1 y_1$$

$$+\int_{t_1^1}^{t_1}\int_{t_1^0}^{y_1}\int_{a}^{t_2}\int_{a}^{b}G(z_2,s_2)\bigg(|p(s_2)|\int_{t_1^1}^{z_1}|u_1(\tau_0(s_1),s_2)|\Delta_1 s_1$$

$$+|u(z_1,s_2)|+|u(t_1^1,s_2)|$$

$$+\sum_{j=0}^{n}\int_{t_1^1}^{z_1}|b_j(s_1)||u_1(\tau_j(s_1),s_2)|\Delta_1 s_1$$

$$+\int_{t_1^1}^{z_1}|f(s_1,s_2,u(s_1,s_2),u_1(\tau_0(s_1),s_2),\ldots,u_1(\tau_n(s_1),s_2))|\Delta_1 s_1\Bigg)$$

$$\Delta_2 s_2 \Delta_2 z_2 \Delta_1 z_1 \Delta_1 y_1\Bigg)$$

$$\leq \epsilon_2 \bigg(A_2 + A_2(t_1^2 - t_1^1)^3(b-a) + A_2^3(t_1^2 - t_1^1)^3(b-a)^2$$

$$+ 2A_2(t_1^2 - t_1^1)^2(b-a)^2 + (n+1)A_2(t_1^2 - t_1^1)^3(b-a)^2$$

$$+ B_2(t_1^2 - t_1^1)^3(b-a)^2 \bigg)$$

$$= \epsilon_2 \bigg(A_2 + A_2(t_1^2 - t_1^1)^3(b-a) + 2A_2(t_1^2 - t_1^1)^2(b-a)^2$$

$$+ (A_2^3 + (n+1)A_2 + B_2)(t_1^2 - t_1^1)^3(b-a)^2 \bigg)$$

$$\leq A_2,$$

$(t_1, t_2) \in [t_1^1, t_1^2] \times [a, b]$. Therefore $T_2(X_2) \subset Y_2$.

Let $z \in T_2(X_2)$ and $y \in Y_2$. Then $\|z\| \leq A_2$ and

$$\|S_2^{-1}y - z\| \leq \|S_2^{-1}y\| + \|z\|$$

$$\leq \frac{\|y\|}{1 + \epsilon_2} + \|z\|$$

$$\leq \frac{C_2}{1 + \epsilon_2} + A_2$$

$$\leq C_2,$$

i.e., $S_2^{-1}y - z \in Y_2$ and

$$\{S_2^{-1}y - z : y \in Y_2, \ z \in T_2(X_2)\} \subset Y_2.$$

Hence and Theorem 17, we conclude that the operator $S_2 + T_2$ has a fixed point $u_2 \in X_2$. From here, $u_2 \in \mathcal{C}_{rd}^1([t_1^1, t_1^2] \times [a, b])$ is a solution of the IBVP

$$u_{t_1}^{\Delta_1}(t_1, t_2) + u_{t_2}^{\Delta_2}(\tau_0(t_1), t_2) + \sum_{j=1}^{n} b_j(t_1) u(\tau_j(t_1), t_2)$$

$$= f(t_1, t_2, u(t_1, t_2), u(\tau_0(t_1), t_2), u(\tau_1(t_1), t_2), \ldots, u(\tau_n(t_1), t_2)),$$

$$t_1 \in [t_1^0, t_1^1], \quad t_2 \in [a, b],$$
$$u(t_1, a) = u(t_1, b), \quad t_1 \in [t_1^1, t_1^2],$$
$$u(t_1, t_2) = u_1(t_1, t_2), \quad t_1 \in [t_1^1, t_1^2], \quad t_2 \in [a, b].$$

3. Continuing this process, we get that the function

$$u(t_1, t_2) = \begin{cases} \phi(t_1, t_2) & \text{if } (t_1, t_2) \in [t_1^{-1}, t_1^0] \times [a, b], \\ u_1(t_1, t_2) & \text{if } (t_1, t_2) \in [t_1^0, t_1^1] \times [a, b], \\ u_2(t_1, t_2) & \text{if } (t_1, t_2) \in [t_1^1, t_1^2] \times [a, b], \\ \vdots & \end{cases}$$

is a solution of the IBVP (7)-(9). This completes the proof.

□

6. Oscillations of First Order Delay Partial Dynamic Equations

Suppose that $\mathbb{T}_1$ is an unbounded above time scale and $\mathbb{T}_2$ is a time scale with forward jump operators and delta differentiation operators σ_1, σ_2 and Δ_1, Δ_2, respectively. Let $t_0 \in \mathbb{T}_1$ and $a_2, b_2 \in \mathbb{T}_2$, $a_2 < b_2$. Consider the BVP for the following first order delay partial dynamic equation.

$$u_{t_1}^{\Delta_1}(t_1, t_2) + u_{t_2}^{\Delta_2}(\tau(t_1), t_2) + b(t_1) u(\tau(t_1), t_2) = 0, \quad t_1 \geq 0, \quad t_2 \in (a_2, b_2), \quad (16)$$

$$u(t_1, a_2) = u(t_1, b_2), \quad t_1 \geq t_0, \quad (17)$$

where

(E1) $b \in \mathcal{C}_{rd}([t_0, \infty))$, $b \geq 0$ on $[t_0, \infty)$,

(E2) $\tau \in \mathcal{C}_{rd}([t_0, \infty))$, $\tau : [t_0, \infty) \to \mathbb{T}_1$, $\tau(t_1) \leq t_1$, $t_1 \in [t_0, \infty)$, $\lim_{t_1 \to \infty} \tau(t_1) = \infty$, $\tau : [t_0, \infty) \to [t_{-1}, \infty)$ is onto, where $t_{-1} = \inf_{t \in [t_0, \infty)} \tau(t)$.

Definition 33. *We say that a solution u of the BVP (16), (17) is eventually positive(negative) if there exists a number $t_1^1 \in [t_0, \infty)$ such that*

$$u(t_1, t_2) > 0 \quad \text{for} \quad \text{any} \quad (t_1, t_2) \in [t_1^1, \infty) \times [a_2, b_2].$$

Definition 34. *A solution u of the BVP (16), (17) is said to be oscillatory if it is neither eventually positive nor eventually negative. Otherwise it is said to be nonoscillatory.*

Definition 35. *The BVP (16), (17) is said to be oscillatory if all its solutions are oscillatory.*

Our first result for the oscillation of the BVP (16), (17) is as follows.

Theorem 19. *Suppose $(E1)$, $(E2)$ and*

$$\liminf_{t \to \infty} \inf_{\lambda \in E} \left(\frac{1}{\lambda e_{-\lambda b}(\tau(t), t)} \right) > 1,$$

where

$$E = \{\lambda \geq 0 : 1 - \lambda b(t) > 0 \quad \text{for} \quad \text{all} \quad \text{large} \quad t \in \mathbb{T}_1\}.$$

Then the BVP (16), (17) is oscillatory.

Proof. Suppose that the BVP (16), (17) has a nonoscillatory solution u. Without loss of generality, assume that u is eventually positive solution of the BVP (16), (17). Then there exists a $t_1^1 \in [t_0, \infty)$ so that

$$u(t_1, t_2) > 0, \quad u(\tau(t_1), t_2) > 0, \quad (t_1, t_2) \in [t_1^1, \infty) \times [a_2, b_2].$$

We integrate both sides of the equation (16) with respect to t_2 from a_2 to b_2, we get

$$\begin{aligned} 0 &= \int_{a_2}^{b_2} u_{t_1}^{\Delta_1}(t_1, s_2) \Delta_2 s_2 + \int_{a_2}^{b_2} u_{t_2}^{\Delta_2}(\tau(t_1), s_2) \Delta_2 s_2 \\ &\quad + b(t_1) \int_{a_2}^{b_2} u(\tau(t_1), s_2) \Delta_2 s_2 \end{aligned}$$

$$= \frac{\partial_1}{\Delta_1 t_1}\left(\int_{a_2}^{b_2} u(t_1,s_2)\Delta_2 s_2\right) + u(\tau(t_1),b_2) - u(\tau(t_1),a_2)$$

$$+ b(t_1)\int_{a_2}^{b_2} u(\tau(t_1),s_2)\Delta_2 s_2$$

$$= \frac{\partial_1}{\Delta_1 t_1}\left(\int_{a_2}^{b_2} u(t_1,s_2)\Delta_2 s_2\right)$$

$$+ b(t_1)\int_{a_2}^{b_2} u(\tau(t_1),s_2)\Delta_2 s_2, \quad t_1 \in [t_1^1,\infty).$$

Let

$$U(t_1) = \int_{a_2}^{b_2} u(t_1,s_2)\Delta_2 s_2, \quad t_1 \in [t_1^1,\infty). \tag{18}$$

Then

$$U(\tau(t_1)) = \int_{a_2}^{b_2} u(\tau(t_1),s_2)\Delta_2 s_2, \quad t_1 \in [t_1^1,\infty),$$

and

$$U^{\Delta_1}(t_1) + b(t_1)U(\tau(t_1)) = 0, \quad t_1 \in [t_1^1,\infty). \tag{19}$$

Since $u(t_1,t_2) > 0$ for any $(t_1,t_2) \in [t_1^1,\infty) \times [a_2,b_2]$, we have that $U(t_1) > 0$ for any $t_1 \in [t_1^1,\infty)$. Therefore the equation (1) has a nonoscillatory solution. This contradicts with Theorem 4. Consequently the BVP (16), (17) is oscillatory. This completes the proof. □

Theorem 20. *Suppose* $(E1)$ *and* $(E2)$. *Assume that there exist* $s \in [t_0,\infty)$ *and* $\lambda_0 \geq 0$ *such that* $-\lambda_0 b \in \mathcal{R}^+([s,\infty))$ *and*

$$\frac{1}{\lambda_0 e_{-\lambda_0 b}(\tau(t),t)} \leq 1$$

for all $t \in [s,\infty)$. *Then the BVP* (16), (17) *has a nonoscillatory solution.*

Proof. Assume that all solutions of the BVP are oscillatory. This contradicts with Theorem 19. Therefore the BVP (16), (17) is nonoscillatory. This completes the proof. □

Theorem 21. *Suppose $(E1)$ and $(E2)$. Assume that τ is eventually nondecreasing and*

$$\limsup_{t\to\infty} \int_{\tau(t)}^{\sigma_1(t)} b(\eta)\Delta_1\eta > 1.$$

Then the BVP (16), (17) is oscillatory.

Proof. Suppose that the BVP (16), (17) has a nonoscillatory solution u. Without loss of generality, assume that u is eventually positive solution of the BVP (16), (17). Then there exists a $t_1^1 \in [t_0, \infty)$ so that

$$u(t_1, t_2) > 0, \quad u(\tau(t_1), t_2) > 0, \quad (t_1, t_2) \in [t_1^1, \infty) \times [a_2, b_2].$$

Let U be defined by (18). Then, as in the proof of Theorem 19, we arrive to the first order dynamic equation (19) for which U is an eventually positive solution. This contradicts with Theorem 6. Consequently the BVP (16), (17) is oscillatory. This completes the proof. $\square$

Theorem 22. *Suppose $(A1)$ and $(A2)$. Assume that τ is eventually nondecreasing and there exists an $\alpha \in (0,1)$ such that*

$$\liminf_{t\to\infty} \int_{\tau(t)}^{t} b(\eta)\Delta_1\eta > \alpha \quad \text{and} \quad \limsup_{t\to\infty} \int_{\tau(t)}^{t} b(\eta)\Delta_1\eta > 1 - \frac{\alpha^2}{4}.$$

Then the BVP (16), (17) is oscillatory.

Proof. Suppose that the BVP (16), (17) has a nonoscillatory solution u. Without loss of generality, assume that u is eventually positive solution of the BVP (16), (17). Then there exists a $t_1^1 \in [t_0, \infty)$ so that

$$u(t_1, t_2) > 0, \quad u(\tau(t_1), t_2) > 0, \quad (t_1, t_2) \in [t_1^1, \infty) \times [a_2, b_2].$$

Let U be defined by (18). Then, as in the proof of Theorem 19, we arrive to the first order dynamic equation (19) for which U is an eventually positive solution. This contradicts with Theorem 7. Consequently the BVP (16), (17) is oscillatory. This completes the proof. $\square$

Theorem 23. *Suppose $(E1)$ and $(E2)$. Assume that τ is eventually nondecreasing and there exists an $\alpha \in (0,1)$ such that*

$$\liminf_{t\to\infty} \int_{\tau(t)}^{t} b(\eta)\Delta_1\eta > \alpha \quad \text{and} \quad \limsup_{t\to\infty} \int_{\tau(t)}^{t} b(\eta)\Delta_1\eta > 1 - \left(1 - \sqrt{1-\alpha}\right)^2.$$

Then the BVP (16), (17) is oscillatory.

Proof. Suppose that the BVP (16), (17) has a nonoscillatory solution u. Without loss of generality, assume that u is eventually positive solution of the BVP (16), (17). Then there exists a $t_1^1 \in [t_0, \infty)$ so that

$$u(t_1, t_2) > 0, \quad u(\tau(t_1), t_2) > 0, \quad (t_1, t_2) \in [t_1^1, \infty) \times [a_2, b_2].$$

Let U be defined by (18). Then, as in the proof of Theorem 19, we arrive to the first order dynamic equation (19) for which U is an eventually positive solution. This contradicts with Theorem 8. Consequently the BVP (16), (17) is oscillatory. This completes the proof. □

Theorem 24. *Suppose $(E1)$ and $(E2)$. Assume that τ is eventually nondecreasing and there exists an $\alpha \in (0, 1)$ such that*

$$\liminf_{t \to \infty} \int_{\tau(t)}^{t} b(\eta) \Delta_1 \eta > \alpha \quad \text{and} \quad \limsup_{t \to \infty} \int_{\tau(t)}^{\sigma(t)} b(\eta) \Delta_1 \eta > 1 - \left(1 - \sqrt{1 - \alpha}\right)^2.$$

Then the BVP (16), (17) is oscillatory.

Proof. Suppose that the BVP (16), (17) has a nonoscillatory solution u. Without loss of generality, assume that u is eventually positive solution of the BVP (16), (17). Then there exists a $t_1^1 \in [t_0, \infty)$ so that

$$u(t_1, t_2) > 0, \quad u(\tau(t_1), t_2) > 0, \quad (t_1, t_2) \in [t_1^1, \infty) \times [a_2, b_2].$$

Let U be defined by (18). Then, as in the proof of Theorem 19, we arrive to the first order dynamic equation (16) for which U is an eventually positive solution. This contradicts with Theorem 9. Consequently the BVP (16), (17) is oscillatory. This completes the proof. □

Theorem 25. *Suppose $(E1)$ and $(E2)$. Assume*

$$\liminf_{t \to \infty} \inf_{\lambda \in [1, \infty)} \left(\frac{1}{\lambda} e_{\lambda b}(\sigma_1(t), \tau(t)) \right) > 1.$$

Then the BVP (16), (17) is oscillatory.

Proof. Suppose that the BVP (16), (17) has a nonoscillatory solution u. Without loss of generality, assume that u is eventually positive solution of the BVP (16), (17). Then there exists a $t_1^1 \in [t_0, \infty)$ so that

$$u(t_1, t_2) > 0, \quad u(\tau(t_1), t_2) > 0, \quad (t_1, t_2) \in [t_1^1, \infty) \times [a_2, b_2].$$

Let U be defined by (18). Then, as in the proof of Theorem 19, we arrive to the first order dynamic equation (19) for which U is an eventually positive solution. This contradicts with Theorem 10. Consequently the BVP (16), (17) is oscillatory. This completes the proof. □

Theorem 26. *Suppose $(E1)$ and $(E2)$. Assume that there exists an $\lambda_0 \in [1, \infty)$ such that*
$$\frac{1}{\lambda_0} e_{\lambda_0 b}(\sigma_1(t), \tau(t)) \leq 1$$
for all large $t \in [t_0, \infty)$. Then the BVP (16), (17) is nonoscillatory.

Proof. Assume that the BVP is oscillatory. This contradicts with Theorem 25. Therefore the BVP (16), (17) is nonoscillatory. This completes the proof. □

Now we consider the following BVP

$$u_{t_1}^{\Delta_1}(t_1, t_2) + u_{t_2}^{\Delta_2}(\tau(t_1), t_2) + \sum_{i=1}^{n} p_i(t_1) u(\tau_i(t_1), t_2) = 0, \quad t_1 \geq 0, \quad t_2 \in (a_2, b_2), \quad (20)$$

$$u(t_1, a_2) = u(t_1, b_2), \quad t_1 \geq t_0, \tag{21}$$

where

(F1) $p_i \in \mathcal{C}_{rd}([t_0, \infty))$, $p_i \geq 0$ on $[t_0, \infty)$, $i \in \{1, \ldots, n\}$,

(F2) $\tau_i \in \mathcal{C}_{rd}([t_0, \infty))$, $\tau_i : [t_0, \infty) \to \mathbb{T}_1$, $\tau_i(t_1) < t_1$, $t_1 \in [t_0, \infty)$, $\lim_{t_1 \to \infty} \tau_i(t_1) = \infty$, $i \in \{0, \ldots, n\}$, where $\tau_0 = \tau$, $\tau_0 : [t_0, \infty) \to [t_{-1}, \infty)$ is onto and $t_{-1} = \inf_{t \in [t_0, \infty)} \tau_0(t)$.

Definition 36. *We say that a solution u of the BVP (20), (21) is eventually positive(negative) if there exists a number $t_1^1 \in [t_0, \infty)$ such that*
$$u(t_1, t_2) > 0 \quad \text{for} \quad \text{any} \quad (t_1, t_2) \in [t_1^1, \infty) \times [a_2, b_2].$$

Definition 37. *A solution u of the BVP (20), (21) is said to be oscillatory if its neither eventually positive nor eventually negative. Otherwise it is said to be nonoscillatory.*

Definition 38. *The BVP (20), (21) is said to be oscillatory if all its solutions are oscillatory.*

Our first result for the oscillation of the BVP (20), (21) is as follows.

Theorem 27. *Suppose* $(F1)$, $(F2)$ *and*

$$\limsup_{t_0 \to \infty} \sup_{t > t_0} \sup_{\lambda \in E} \frac{\lambda \sum_{i=1}^{n} p_i(t)}{\sum_{i=1}^{n} p_i(t) \exp\left(-\int_{\tau_i(t)}^{t} \frac{1}{\mu(s)} \text{Log}\left(1 - \lambda \mu(s) \sum_{i=1}^{n} p_i(s)\right) \Delta_1 s\right)} < 1,$$

where

$$E = \left\{ \lambda > 0 : 1 - \lambda \left(\sum_{i=1}^{n} p_i(t) \right) \mu(t) > 0 \right\},$$

Then the BVP (20), (21) *is oscillatory.*

Proof. Suppose that the BVP (20), (21) has a nonoscillatory solution u. Without loss of generality, assume that u is eventually positive solution of the BVP (20), (21). Then there exists a $t_1^1 \in [t_0, \infty)$ so that

$$u(t_1, t_2) > 0, \quad u(\tau(t_1), t_2) > 0, \quad (t_1, t_2) \in [t_1^1, \infty) \times [a_2, b_2].$$

We integrate both sides of the equation (20) with respect to t_2 from a_2 to b_2, we get

$$\begin{aligned}
0 &= \int_{a_2}^{b_2} u_{t_1}^{\Delta_1}(t_1, s_2) \Delta_2 s_2 + \int_{a_2}^{b_2} u_{t_2}^{\Delta_2}(\tau(t_1), s_2) \Delta_2 s_2 \\
&\quad + \sum_{i=1}^{n} p_i(t_1) \int_{a_2}^{b_2} u(\tau_i(t_1), s_2) \Delta_2 s_2 \\
&= \frac{\partial_1}{\Delta_1 t_1} \left(\int_{a_2}^{b_2} u(t_1, s_2) \Delta_2 s_2 \right) + (u(\tau(t_1), b_2) - u(\tau(t_1), a_2)) \\
&\quad + \sum_{i=1}^{n} p_i(t_1) \int_{a_2}^{b_2} u(\tau_i(t_1), s_2) \Delta_2 s_2 \\
&= \frac{\partial_1}{\Delta_1 t_1} \left(\int_{a_2}^{b_2} u(t_1, s_2) \Delta_2 s_2 \right) \\
&\quad + \sum_{i=1}^{n} p_i(t_1) \int_{a_2}^{b_2} u(\tau_i(t_1), s_2) \Delta_2 s_2, \quad t_1 \in [t_1^1, \infty).
\end{aligned}$$

Let

$$U(t_1) = \int_{a_2}^{b_2} u(t_1, s_2) \Delta_2 s_2, \quad t_1 \in [t_1^1, \infty).$$

Then
$$U(\tau(t_1)) = \int_{a_2}^{b_2} u(\tau(t_1), s_2)\Delta_2 s_2, \quad t_1 \in [t_1^1, \infty),$$
and
$$U^{\Delta_1}(t_1) + \sum_{i=1}^{n} p_i(t_1) U(\tau_i(t_1)) = 0, \quad t_1 \in [t_1^1, \infty). \tag{22}$$

Since $u(t_1, t_2) > 0$ for any $(t_1, t_2) \in [t_1^1, \infty) \times [a_2, b_2]$, we have that $U(t_1) > 0$ for any $t_1 \in [t_1^1, \infty)$. Therefore the equation (22) has a nonoscillatory solutiuon. This contradicts with Theorem 12. Consequently the BVP (20), (21) is oscillatory. This completes the proof. $\square$

REFERENCES

[1] Hilger S., Analysis on measure chains: A unified approach to continuous and discrete calculus, *Results Math.* 18 (1990) 18-56.

[2] Bohner M. and Peterson A., *Dynamic Equations on Time Scales: An Introduction with Applications*, Birkhäuser, Boston, 2003.

[3] Georgiev S., *Integral Equations on Time Scales*. Atlantis Press 2016.

[4] Bohner M. and Georgiev S., *Multidimensional Time Scale Calculus*. Springer 2016.

[5] Zhang B. and Deng X., Oscillation of Delay Differential Equations on Time Scales, *Math. Comput. Modelling*, 2002, 36(11-13), pp. 1307-1318.

[6] Bohner M., Some Oscillation Criteria for First Order Delay Dynamic Equations, *Far East J. Appl. Math.*, 2005, 18(3), pp. 289-304.

[7] Bohner M., Karpuz B. and Ocalan O., Iterated Oscillation Criteria for Delay Dynamic Equations of First Order, *Adv. Difference Equ.*, 2008.

[8] Braverman E. and Karpuz B., Nonoscillation of First Order Dynamic Equations with Several Delays, *Adv. Difference Equ.*, 2010.

[9] Sahiner Y. and Stavroulakis I., Oscillations of First Order Delay Dynamic Equations, *Dynam. Systems Appl.*, 2006, 15(3-4), pp. 645-655.

[10] Agarwal R. and Bohner M., An Oscillation Criterion for First Order Delay Dynamic Equations, *Funct. Differ. Equ.*, 2009. 16(1), pp. 11-17.

[11] Karpuz B., *Sharp Oscillation and Nonoscillation Tests for Dynamic Delay Equations*, MMAS, in press.

[12] Agwo A., On the Oscillation of First Order Delay Dynamic Equations with Variable Coefficients, *Rocky Mountain Journal of Mathematics*, Vol. 38, N. 1, 2008.

[13] Braverman E. and Karpuz B., Uniform Expoenential Stability of First Order Dynamic Equations with Several Delays, *Applied Mathematics and Computation*, 218(2012), pp. 10468-10485.

[14] Anderson D., Global Stability for Nonlinear Dynamic Equations with Variable Coefficients, *J. Math. Anal. Appl.*, 345(2008), pp. 796-804.

[15] Xiang T. and Yuan R., A class of expansive type Krasnoselskii fixed point theorem. *Nonlinear Analysis*, 2009, pp. 3229 - 3239.

In: Asymptotic Behavior: An Overview
Editor: Steve P. Riley

ISBN: 978-1-53617-222-5
© 2020 Nova Science Publishers, Inc.

Chapter 2

ASYMPTOTIC BEHAVIOR IN QUANTUM-FIELD MODELS FROM SCHWINGER-DYSON EQUATIONS

V. E. Rochev[*]

Division of Theoretical Physics,
A. A. Logunov Institute for High Energy Physics,
NRC "Kurchatov Institute", Protvino, Russia

Abstract

Asymptotic behavior in the deep Euclidean region of momenta for four-dimensional models of quantum field theory is studied by using the system of Schwinger-Dyson equations (SDEs). This system is truncated by a sequence of n-particle approximations which for $n \to \infty$ goes into the complete system of SDEs. For a model of complex scalar field ϕ with interaction $\frac{\lambda}{2}(\phi^*\phi)^2$ an asymptotic solution of the system of SDEs in two-particle approximation is obtained. The two-particle amplitude has the pathology-free asymptotic behavior at large momenta in the region of strong coupling at $\lambda > \lambda_{cr}$. At $\lambda < \lambda_{cr}$ this amplitude possesses Landau-type singularity. For pseudoscalar Yukawa model a solution of the two-particle approximation for the pseudoscalar propagator is free from non-physical singularities and has the self-consistent asymptotic behavior. The investigation of the super-renormalized model of a complex

---

[*]Corresponding Author's E-mail: rochev@ihep.ru.

scalar field ϕ and a real scalar field χ with the interaction $g\phi^*\phi\chi$ demonstrates a change of the asymptotic behavior in the Euclidean region of momenta in a vicinity of a certain critical value of the coupling constant. For small values of the coupling the propagator of field ϕ behaves asymptotically as free. In the strong-coupling region the asymptotic behavior drastically changes – the propagator in the deep Euclidean region tends to a constant. In the coordinate space this propagator has a characteristic shell structure. The same shell structure in coordinate space has a vertex with zero momentum transfer. An analogy between the phase transition in this model and the re-arrangement of the physical vacuum in the supercritical external field due to the "fall-on-the-center" phenomenon is discussed.

PACS: 11.10.Jj

Keywords: quantum field theory, Schwinger-Dyson equations, asymptotic behavior

1. INTRODUCTION

Fundamental particles and their interactions are successfully described by the Standard Model (SM). Nevertheless, SM has some essential problems[1]. One of the main problems of SM is Landau poles for the Abelian hypercharge coupling and in the Higgs-Yukawa sector. A theory with the Landau pole cannot be self consistent, since the Landau pole has a wrong sign residue that indicates the presence of unphysical ghost fields, which leads to the violation of causality.

The Landau pole was first discovered by Landau and coworkers [4, 5] in an attempt to determine the asymptotic behavior of the photon propagator at high momenta. Their investigation was based on the approximate solution of Dyson equations with the summation of leading logarithms. Similar poles were indicated also in other strictly renormalized models: self-interacting scalar field and Yukawa interaction. Such singularities in Euclidean region of momenta violate general principles of the local field theory and is a serious problem of quantum field theory [6].

Further development has demonstrated that these non-physical singularities arise practically inevitably in the framework of any well-known methods: at the renormalization-group summation, in the frameworks of $1/N$-expansion and

---
[1] See recent reviews [1]– [3] of SM and its problems.

mean-field expansion, etc. A widespread opinion is formulated as a triviality of the quantum field models that is not asymptotically free [2]. There is a rigorous theorem [10] – [12] that the four-dimensional scalar field theory with ϕ^4 interaction on the lattice does not have an interacting continuum theory as its limit for zero lattice spacing, i.e., the theory is trivial [3].

In recent years, a new direction has been outlined in the fight against the problem of triviality in SM (see [14], [15] for review and further references). This direction is based on Weinberg's old idea [16], [17] of asymptotic safety as a possible panacea for working with non-renormalizable and non-asymptotically-free interactions. Asymptotic safety generalizes asymptotic freedom and it is a candidate scenario to provide an ultraviolet extension for the quantum field theory of gravity through an interacting fixed point of the renormalization group (see [18] for review). Moreover, an asymptotically safe fixed point might provide the quantum-gravity induced UV-completion of the SM and could offer a mechanism to solve the Landau-pole problem. However, there are reasonable doubts that this program can be implemented without a radical departure from the framework of perturbation theory (see [19], [20]).

Therefore, it seems inevitable to conclude that solving the Landau-pole problem requires going beyond the framework of the coupling–constant perturbation theory.

A first attempt to remove these non-physical singularities was made in works [21], [22]. The method of these works was based on the application of Källén–Lehmann representation to restore the correct analytical structure. This method remains the main way to solve the problem of the non-physical singularities up to our time. However, the absence of dynamical foundation required a more detailed investigation of non-perturbative region.

Without claiming to have a complete bibliography of works in this direction, we point out Suslov's series of works ([23] – [25]; for a complete reference list see [26]) which substantiates the linear dependence of the Gell-Mann – Low function on the coupling constant in the strong coupling region for the scalar ϕ^4-theory and quantum electrodynamics. Such a dependence solves the Landau-pole problem and the triviality problem [4]. Other works in this direction

---

[2] See also [7]–[9] for the historical survey and further references.

[3] However, as Weinberg points out [13], this argument is not fully conclusive due to an uncertainty of the continuous limit in this model.

[4] It should be noted that the Suslov's method of researching was sharply criticized in works [27], [28].

is usually offer a solution to the problem beyond traditional models (see, e.g. [29], [30]).

The asymptotic short-distance region in the models without asymptotic freedom is the region of non-weak coupling, and it is the main difficulty in its investigation. It seems that the standard methods (renormalization-group summation, $1/N$-expansion etc.) are too tethered to the weak-coupling region and are not enough meaningful to inform us about behavior at short distances in these models. In any case, a study of the Landau-pole problem requires a non-perturbative tool.

In this chapter we overview the study of the problem of asymptotic behavior in models of quantum field theory by the Schwinger-Dyson equation method [31]–[36].

For the model of a self-acting scalar field and the Yukawa model we propose a new non-perturbative approximation – two-particle approximation. This approximation is based on a system of Schwinger-Dyson equations (SDEs) for the propagator and the two-particle function. The first equation is the exact SDE equation for the propagator, and the second equation is a truncated SDE for the two-particle function. The total system of SDEs is, in fact, the system of relations between derivatives of the generating functional of correlation functions, resulting from the functional-differential SDE, which acts as a dynamical principle of the theory. If we approximate the generating functional with the first n terms of an expansion in powers of a source, the system of SDEs can be approximated by a closed system of integral equations. This system defines the n-th term of a sequence of approximations, which for $n \to \infty$ obviously goes into the complete system of SDEs. The two-particle approximation is considered as a first non-trivial step of the sequence of general n-particle approximations, which tends to the exact infinite system of SDEs at $n \to \infty$. An investigation of some truncation of SDEs is not a news, of course. A new step is the consideration of a system of these equations instead of the usual consideration of some single truncated equation.

We will also discuss in this chapter the asymptotic behavior in the Yukawa scalar model, i.e., a model of a complex scalar field ϕ (phion) and a real field χ (chion) with the interaction $g\phi^*\phi\chi$. This model is used in nuclear physics as a simplified version of the Yukawa model without spin degrees of freedom, as well as an effective model of the interaction of scalar quarks [37], [38]. Despite its well-known imperfection associated with its instability [39] (or more precisely, the metastability [40] – [41]), this model, as the simplest four-dimensional model of the interaction of fields, often used as a prototype of more substantive theories to elaborate the various non-perturbative approaches in the quantum field theory.

A structure of the chapter is the following: In Section 2 a general construction of the approximation schemes for the system of SDEs is considered: the necessary notations and definitions are given and SDE for the generating functional of correlation functions are introduced in the formalism of a bilocal source[5]. We consider using of the bilocal source as a convenient choice of the functional variable.

In Section 3 the problem of asymptotic behavior in renormalizable models is investigated.

For a model of a complex scalar field with self-action in four dimensions (Subsection 3.1) the asymptotic solution of the equation for the two-particle amplitude at large momenta is presented and the asymptotic behavior of the amplitude at large momenta is discussed. The amplitude possesses in this model a self-consistent behavior (as a constant plus a decreasing oscillating term) at the values of the renormalized coupling $\lambda > \lambda_{cr}$. At $\lambda < \lambda_{cr}$ the amplitude has some Landau-type singularity.

For Yukawa model (Subsection 3.2) the asymptotic solution of the system of two-particle approximation at large Euclidean momenta is presented and the asymptotic behavior of the propagators at large momenta is discussed. The boson propagator in this model possesses self-consistent behavior.

In Section 4 we consider scalar Yukawa model in four dimensions. The study of this model shows a change of the asymptotic behavior of propagators in the deep Euclidean region in a vicinity of a certain critical value of the coupling constant. For small values of the coupling the propagators behave as free, which is consistent with the wide-spread opinion about the dominance of perturbation theory for this super-renormalizable model. In the strong-coupling region, however, the asymptotic behavior changes dramatically – the phion propagator in the deep Euclidean region tend to some constant limits.

The existence of a critical coupling constant in the scalar Yukawa model was noticed by practically all authors who have investigated this model using different methods (see, e.g., [47],[48] and references therein). This critical constant is usually regarded as a limit on the coupling constant for a self-consistent description of the model by some method. In our approach, however, the self-

---

[5]A formalism of the bilocal source was first elaborated in QFT by Dahmen and Jona-Lasinio [42] (see also [43]– [46]).

consistent solution for propagators exists also for the strong coupling, and the existence of the critical coupling looks more like as a phase transition in accordance with the general definition of the phase transition as a sharp change of properties of the model with a smooth change of parameters (see, e.g., [49]).

2. GENERAL CONSIDERATION: SCHWINGER-DYSON EQUATIONS AND APPROXIMATION SCHEMES

2.1. System of Schwinger-Dyson Equations

We consider the model of interaction of a complex scalar field ϕ (phion) and a real scalar field χ (chion) with the Lagrangian

$$\mathcal{L} = -\partial_\mu \phi^* \partial_\mu \phi - m_0^2 \phi^* \phi - \frac{1}{2}(\partial_\mu \chi)^2 - \frac{\mu_0^2}{2}\chi^2 + g\phi^* \phi \chi \quad (1)$$

in a d-dimensional Euclidean space ($x \in E_d$). Here m_0 and μ_0 are bare masses.

The generating functional of correlation functions (vacuum averages) is the functional integral

$$G(\eta, j) = \int D(\phi, \phi^*, \chi) \exp\left\{ \int dx\, \mathcal{L}(x) - \int dx dy\, \phi^*(y)\eta(y,x)\phi(x) + \int dx\, j(x)\chi(x) \right\}. \quad (2)$$

Here η is a bilocal source of the phion field, j is a single source of the chion field.

The translational invariance of the functional integration measure leads to relations

$$\int D(\phi, \phi^*, \chi) \frac{\delta}{\delta \phi^*(x)} \phi^*(y) \exp\left\{ \int dx\, \mathcal{L}(x) - \int dx dy\, \phi^*(x)\eta(x,y)\phi(y) + \int dz\, j(z)\chi(z) \right\} = 0,$$

and

$$\int D(\phi, \phi^*, \chi) \frac{\delta}{\delta \chi(z)} \exp\left\{ \int dx\, \mathcal{L}(x) - \int dx dy\, \phi^*(x)\eta(x,y)\phi(y) + \int dz\, j(z)\chi(z) \right\} = 0.$$

Since $\delta G/\delta \eta(y,x) = - <\phi(x)\phi^*(y)>$ and $\delta G/\delta j(z) = <\chi(z)>$ we can rewrite these relations as the functional-differential Schwinger-Dyson equations (SDEs) for generating functional G:

$$g\frac{\delta^2 G}{\delta\eta(y,x)\delta j(x)} = (m_0^2 - \partial_x^2)\frac{\delta G}{\delta\eta(y,x)} + \int dx_1 \eta(x,x_1)\frac{\delta G}{\delta\eta(y,x_1)} + \delta(x-y)G \quad (3)$$

and

$$g\frac{\delta G}{\delta\eta(z,z)} + (\mu_0^2 - \partial^2)\frac{\delta G}{\delta j(z)} = j(z)G. \quad (4)$$

In terms of the logarithm $Z = \log G$ these equations are

$$g\left[\frac{\delta^2 Z}{\delta\eta(y,x)\delta j(x)} + \frac{\delta Z}{\delta j(x)}\frac{\delta Z}{\delta\eta(y,x)}\right] = (m_0^2 - \partial_x^2)\frac{\delta Z}{\delta\eta(y,x)} + \int dx_1\, \eta(x,x_1)\frac{\delta Z}{\delta\eta(y,x_1)} + \delta(x-y), \quad (5)$$

$$\frac{\delta Z}{\delta j(x)} = \int dx_1\, D_c(x-x_1)\, j(x_1) - g\int dx_1\, D_c(x-x_1)\frac{\delta Z}{\delta\eta(x_1,x_1)}. \quad (6)$$

Here $D_c \equiv (\mu_0^2 - \partial^2)^{-1}$ is a free chion propagator.

Equation (6) allows us to express all correlation functions with chion legs in terms of functions that contain phions only. Thus, the differentiation of (6) over η gives us the three-point function

$$V(x,y|z) \equiv -\frac{\delta^2 Z}{\delta j(z)\delta\eta(y,x)}\bigg|_{\eta=j=0} = g\int dz_1\, D_c(z-z_1)\, Z_2\begin{pmatrix} z_1 & x \\ z_1 & y \end{pmatrix}, \quad (7)$$

where

$$Z_2\begin{pmatrix} x & x' \\ y & y' \end{pmatrix} \equiv \frac{\delta^2 Z}{\delta\eta(y',x')\delta\eta(y,x)}\bigg|_{\eta=j=0} \quad (8)$$

is the two-particle phion function. The differentiation of (6) over j with taking into account equation (7) gives us the chion propagator:

$$D(x-y) \equiv \frac{\delta^2 Z}{\delta j(y)\delta j(x)}\bigg|_{\eta=j=0} = D_c(x-y) + \int dx_1 dy_1\, \alpha(x-x_1) Z_2\begin{pmatrix} x_1 & y_1 \\ x_1 & y_1 \end{pmatrix} D_c(y_1-y). \quad (9)$$

Here

$$\alpha(x) = g^2 D_c(x).$$

We can exclude with the help of the SDE (6) a differentiation over j in SDE (5). Then we obtain at $j=0$ the SDE for generating functional Z:

$$\int dx_1\, \alpha(x-x_1)\left[\frac{\delta^2 Z}{\delta\eta(x_1,x_1)\delta\eta(y,x)} + \frac{\delta Z}{\delta\eta(x_1,x_1)}\frac{\delta Z}{\delta\eta(y,x)}\right] +$$

$$+(m_0^2 - \partial_x^2)\frac{\delta Z}{\delta\eta(y,x)} + \int dy_1\, \eta(x,y_1)\frac{\delta Z}{\delta\eta(y,y_1)} + \delta(x-y) = 0. \quad (10)$$

Equation (10) only contains the derivatives over the bilocal source η. Note that if we make the change

$$\alpha(x) \to -\lambda\delta(x) \qquad (11)$$

in this equation, the resulting equation will exactly coincide with the SDE for the theory of complex scalar fields with self-action

$$\mathcal{L}_{int} = -\frac{\lambda}{2}(\phi^*\phi)^2 \qquad (12)$$

(see [31]). Thus, all of the following general constructions are valid both for the theory of the interaction of phions with chion and for the theory of complex scalar field with the self-action.

Differentiations of this equation give us the infinite system of SDEs for correlation functions. We write the first three equations of this system which are necessary for our following constructions. Switching off the source in equation (10), we obtain the equation

$$(m^2 - \partial_x^2)\Delta(x-y) = \delta(x-y) + \int dx_1\, \alpha(x-x_1)\, Z_2\begin{pmatrix} x & x_1 \\ y & x_1 \end{pmatrix}. \qquad (13)$$

Here

$$\Delta(x-y) \equiv -\left.\frac{\delta Z}{\delta \eta(y,x)}\right|_{\eta=0} \qquad (14)$$

is the phion propagator,

$$m^2 \equiv m_0^2 - \alpha(p=0)\Delta(x=0), \qquad (15)$$

and

$$\alpha(p) = \int dx\, e^{ipx}\alpha(x)$$

is a Fourier image of $\alpha(x)$.

First differentiation over η gives us the equation:

$$(m^2 - \partial_x^2)Z_2\begin{pmatrix} x & x' \\ y & y' \end{pmatrix} - \int dx_1\, \alpha(x-x_1)\, Z_2\begin{pmatrix} x_1 & x' \\ x_1 & y' \end{pmatrix}\Delta(x-y) +$$

$$+ \int dx_1\, \alpha(x-x_1)\, Z_3\begin{pmatrix} x_1 & x & x' \\ x_1 & y & y' \end{pmatrix} = \delta(x-y')\Delta(x'-y). \qquad (16)$$

The second differentiation over η gives one more equation:

$$(m^2 - \partial_x^2) Z_3 \begin{pmatrix} x & x' & x'' \\ y & y' & y'' \end{pmatrix} - \int dx_1\, \alpha(x-x_1)\, Z_3 \begin{pmatrix} x_1 & x' & x'' \\ x_1 & y' & y'' \end{pmatrix} \Delta(x-y) +$$
$$+ \int dx_1\, \alpha(x-x_1)\, Z_4 \begin{pmatrix} x_1 & x & x' & x'' \\ x_1 & y & y' & y'' \end{pmatrix} = -\left\{ \delta(x-y'') Z_2 \begin{pmatrix} x' & x'' \\ y' & y \end{pmatrix} + \right.$$
$$\left. + \int dx_1\, \alpha(x-x_1)\, Z_2 \begin{pmatrix} x_1 & x' \\ x_1 & y' \end{pmatrix} Z_2 \begin{pmatrix} x'' & x \\ y'' & y \end{pmatrix} \right\} - \left\{ x' \leftrightarrow x'', y' \leftrightarrow y'' \right\}. \quad (17)$$

Here $Z_n \equiv \left. \dfrac{\delta^n Z}{\delta \eta^n} \right|_{\eta=0}$ is the n-particle phion function.

A system of SDEs is the infinite set of integral equations containing, in principle, all the information about a model of quantum field theory. So far, there are no effective methods for the study of this infinite system as a whole, so it is necessary to truncate this system with a finite set of equations. This truncation usually refers to either an expansion in a small parameter (examples of such expansions are the coupling-constant perturbation theory and the $1/N$-expansion), or a simulation (and rather, guessing) of some properties of the model. An initial truncation of the system determines further approximations of the complete system, which is thus attached to the leading approximation.

2.2. Mean-Field Expansion

As the leading approximation for the mean-field expansion we take the equation

$$\int dx_1\, \alpha(x-x_1)\, \frac{\delta Z^{(0)}}{\delta \eta(x_1, x_1)} \frac{\delta Z^{(0)}}{\delta \eta(y, x)} + (m_0^2 - \partial_x^2) \frac{\delta Z^{(0)}}{\delta \eta(y, x)} + \int dy_1\, \eta(x, y_1)\, \frac{\delta Z^{(0)}}{\delta \eta(y, y_1)} + \delta(x-y) = 0, \quad (18)$$

where $Z^{(0)}$ is the leading-order generating functional.

In correspondence with equation (10) the next-to-the-leading (NLO) equation will be

$$\int dx_1\, \alpha(x-x_1)\, \frac{\delta Z^{(1)}}{\delta \eta(x_1, x_1)} \frac{\delta Z^{(0)}}{\delta \eta(y, x)} + \int dx_1\, \alpha(x-x_1)\, \frac{\delta Z^{(0)}}{\delta \eta(x_1, x_1)} \frac{\delta Z^{(1)}}{\delta \eta(y, x)} +$$
$$+ (m_0^2 - \partial_x^2) \frac{\delta Z^{(1)}}{\delta \eta(y, x)} + \int dy_1\, \eta(x, y_1)\, \frac{\delta Z^{(1)}}{\delta \eta(y, y_1)} = - \int dx_1\, \alpha(x-x_1)\, \frac{\delta^2 Z^{(0)}}{\delta \eta(x_1, x_1) \delta \eta(y, x)}. \quad (19)$$

After switching off the source equation (18) gives us the leading-order phion propagator Δ_0. In the momentum space

$$\Delta_0^{-1}(p) = m^2 + p^2, \quad (20)$$

where m^2 is given be equation (15).

Linear integral equations

$$\int dx_1 dy_1 K_{12}\begin{pmatrix} x & x_1 \\ y & y_1 \end{pmatrix} F\begin{pmatrix} x_1 & x' \\ y_1 & y' \end{pmatrix} = F_0\begin{pmatrix} x & x' \\ y & y' \end{pmatrix}$$

with the kernel

$$K_{12}\begin{pmatrix} x & x' \\ y & y' \end{pmatrix} = \delta(x-x')\delta(y-y') - \delta(x'-y')\int dx_1 \Delta_1(x-x_1)\alpha(x_1-y')\Delta_2(x_1-y) \quad (21)$$

will be repeatedly considered below. Here Δ_1 and Δ_2 are the given functions. It is easy to verify that the inverse kernel K_{12}^{-1}, which is defined as

$$\int dx_1 dy_1 K_{12}^{-1}\begin{pmatrix} x & x_1 \\ y & y_1 \end{pmatrix} K_{12}\begin{pmatrix} x_1 & x' \\ y_1 & y' \end{pmatrix} = \delta(x-x')\delta(y-y'), \quad (22)$$

has the form

$$K_{12}^{-1}\begin{pmatrix} x & x' \\ y & y' \end{pmatrix} = \delta(x-x')\delta(y-y') - \delta(x'-y')\int dx_1 \Delta_1(x-x_1) f(x_1-y')\Delta_2(x_1-y), \quad (23)$$

where $f(x)$ is a solution of equation

$$f(x) = -\alpha(x) + \int dx_1 dy_1 f(x-x_1) L_{12}(x_1-y_1)\alpha(y_1). \quad (24)$$

Here

$$L_{12}(x) = \Delta_1(x)\Delta_2(-x). \quad (25)$$

This equation can be easily solved by transition to the momentum space, and we have

$$\frac{1}{f(p)} = -\frac{1}{\alpha(p)} + L_{12}(p), \quad (26)$$

where $f(p)$ and $L(p)$ are Fourier images of $f(x)$ and $L_{12}(x)$.

In particular, a differentiation of the leading-order equation of mean-field expansion over η gives us the equation for the two-particle function $Z_2^{(0)}$, which can be rewritten as

$$\int dx_1 dy_1 K_0\begin{pmatrix} x & x_1 \\ x & y_1 \end{pmatrix} Z_2^{(0)}\begin{pmatrix} x_1 & x' \\ y_1 & y' \end{pmatrix} = \Delta_0(x-y')\Delta_0(x'-y), \quad (27)$$

where K_0 is kernel (21) at $\Delta_1 = \Delta_2 = \Delta_0$. In correspondence with equations (23)–(26), the solution of this equation is

$$Z_2^{(0)}\begin{pmatrix} x & x' \\ y & y' \end{pmatrix} = \Delta_0(x-y')\Delta_0(x'-y) +$$

$$-\int dx_1 dx_2 \Delta_0(x-x_1)\Delta_0(x'-x_2)f_0(x_1-x_2)\Delta_0(x_1-y)\Delta_0(x_2-y'), \quad (28)$$

where

$$f_0^{-1} = -\alpha^{-1} + L_0, \quad (29)$$

and $L_0(x) = \Delta_0(x)\Delta_0(-x)$ is the scalar loop.

In correspondence with equation (9), the leading-order chion propagator in the momentum space will be

$$D_0^{-1}(p) = \mu_0^2 + p^2 - g^2 L_0(p). \quad (30)$$

The repeated differentiation leads to the equation for the leading-order three-particle function whose solution is

$$Z_3^{(0)}\begin{pmatrix} x & x' & x'' \\ y & y' & y'' \end{pmatrix} = \int dx_1 dy_1\, K_0^{-1}\begin{pmatrix} x & x_1 \\ y & y_1 \end{pmatrix} Z_{30}\begin{pmatrix} x_1 & x' & x'' \\ y_1 & y' & y'' \end{pmatrix}. \quad (31)$$

Here

$$Z_{30}\begin{pmatrix} x & x' & x'' \\ y & y' & y'' \end{pmatrix} \equiv -\left\{\delta(x-y')Z_2^{(0)}\begin{pmatrix} x' & x'' \\ y & y'' \end{pmatrix} + \right.$$

$$\left. + \int dx_1 \alpha(x-x_1)\, Z_2^{(0)}\begin{pmatrix} x_1 & x' \\ x_1 & y' \end{pmatrix} Z_2^{(0)}\begin{pmatrix} x & x'' \\ y & y'' \end{pmatrix}\right\} - \left\{x' \leftrightarrow x'', y' \leftrightarrow y''\right\}. \quad (32)$$

The following property of the three-particle function

$$\int dx_1\, \alpha(x-x_1) Z_3^{(0)}\begin{pmatrix} x_1 & x & x' \\ x_1 & y & y' \end{pmatrix} = -\int dx_1\, f_0(x-x_1)\, Z_{30}\begin{pmatrix} x_1 & x & x' \\ x_1 & y & y' \end{pmatrix} \quad (33)$$

can be easy derived from above formulas.

The other functions of the leading approximation can be calculated in the same manner.

Note that in contrast to the functional derivatives of Z over single sources, the higher derivatives of Z over bilocal source η are not connected parts of corresponding many-particle functions. Thus, the two-particle phion function

Z_2 is not the connected part Z_2^c of two-particle function and related to it by the formula:

$$Z_2\begin{pmatrix} x & x' \\ y & y' \end{pmatrix} = \Delta(x-y')\Delta(x'-y) + Z_2^c\begin{pmatrix} x & x' \\ y & y' \end{pmatrix}. \quad (34)$$

A characteristic feature of many-particle functions of the leading mean-field approximation is their incomplete structure in terms of crossing symmetry. The Bose symmetry of the theory dictates the crossing symmetry of the connected part Z_2^c:

$$Z_2^c\begin{pmatrix} x & x' \\ y & y' \end{pmatrix} = Z_2^c\begin{pmatrix} x' & x \\ y' & y \end{pmatrix} = Z_2^c\begin{pmatrix} x' & x \\ y & y' \end{pmatrix} = Z_2^c\begin{pmatrix} x & x' \\ y' & y \end{pmatrix}. \quad (35)$$

It is easy to see that $Z_2^{(0)c}$ (this is the second term on the rhs of equation (28)) satisfies the first equality in (35), but breaks the other two. This apparent discrepancy is a feature of many non-perturbative approximations. It is inherent, for example, to the Bethe-Salpeter equation in the ladder approximation. We face similar problems in the two-particle approximation, which will be considered below. To restore the missing crossing symmetry of the leading approximation, it is necessary to look at the next order.

Calculations in the next order quite similar to the above. Equation (19) with the source being switched off gives us the equation for the first correction Δ_1 to the propagator. The differentiation of equation (19) over the source gives the equation for the NLO two-particle function $Z_2^{(1)}$, which can be written as an equation with kernel K_0, and taking into account the above formulae the solution is

$$Z_2^{(1)}\begin{pmatrix} x & x' \\ y & y' \end{pmatrix} = Z_{21}\begin{pmatrix} x & x' \\ y & y' \end{pmatrix} - \int dx_1 dx_2 \Delta_0(x-x_1) f_0(x_1-x_2) \Delta_0(x_1-y) Z_{21}\begin{pmatrix} x_2 & x' \\ x_2 & y' \end{pmatrix}, \quad (36)$$

where

$$Z_{21}\begin{pmatrix} x & x' \\ y & y' \end{pmatrix} \equiv -\int dx_1 dy_1 \Delta_0(x-y_1)\alpha(y_1-x_1) Z_3^{(0)}\begin{pmatrix} x_1 & y_1 & x' \\ x_1 & y_1 & y' \end{pmatrix} +$$

$$+\Delta_0(x-y')\Delta_1(x'-y) - \alpha(p=0)\Delta_1(x=0)\int dx_1 \Delta_0(x-x_1) Z_2^{(0)}\begin{pmatrix} x_1 & x' \\ y & y' \end{pmatrix} +$$

$$+\int dx_1 dx_2 \Delta_0(x-x_1)\alpha(x_1-x_2)\Delta_1(x_1-y) Z_2^{(0)}\begin{pmatrix} x_2 & x' \\ x_2 & y' \end{pmatrix}. \quad (37)$$

From these equations, together with equation (33), it follows that $Z_2^{(1)}$ contains the term

$$-\int dx_1 dx_2 \Delta_0(x-x_1)\Delta_0(x'-x_2)f_0(x_1-x_2)\Delta_0(x_2-y)\Delta_0(x_1-y') =$$
$$= Z_2^{(0)c}\begin{pmatrix} x & x' \\ y' & y \end{pmatrix} = Z_2^{(0)c}\begin{pmatrix} x' & x \\ y & y' \end{pmatrix}, \quad (38)$$

which restores the missing crossing symmetry of the leading-order two-particle function. Such restoration of the crossing symmetry is typical for non-perturbative expansions in the formalism of bilocal source (for similar examples in other models see [46], [51]).

2.3. $1/N$-Expansion in the Framework of Bilocal-Source Formalism

Although the technique of construction of the $1/N$ – expansion for vector-matrix models is well-known (see [52] for the diagrammatic method and [53] for the functional method), we include in our review, for the sake of completeness, a brief summary of a construction of this expansion in the framework of bilocal-source formalism.

The Lagrangian of the model is a N-component generalization of Lagrangian (1):

$$\mathcal{L} = -\partial_\mu \phi_a^* \partial_\mu \phi^a - m_0^2 \phi_a^* \phi^a - \frac{1}{2}(\partial_\mu \chi_b^a)^2 - \frac{\mu_0^2}{2}(\chi_b^a)^2 + \frac{g}{\sqrt{N}}\phi_a^* \phi^b \chi_b^a, \quad (39)$$

where $a, b = 1, \cdots, N$. The usual agreement on the summation of repeated discrete indices is implied. Introducing bilocal matrix source $\eta_b^a(x,y)$ for phions and single matrix source $j_b^a(x)$ for chion it is easy to obtain as a direct generalization of results of Subsection 2.1 the system of SDEs for this model.

$$\frac{g}{\sqrt{N}}\frac{\delta^2 G}{\delta \eta_{b'}^b(y,x)\delta j_a^{b'}(x)} = (m_0^2-\partial_x^2)\frac{\delta G}{\delta \eta_a^b(y,x)} + \int dx_1 \eta_{a'}^a(x,x_1)\frac{\delta G}{\delta \eta_{a'}^b(y,x_1)} + \delta_b^a \delta(x-y)G,$$
$$(40)$$

$$\frac{g}{\sqrt{N}}\frac{\delta G}{\delta \eta_a^b(z,z)} + (\mu_0^2-\partial^2)\frac{\delta G}{\delta j_a^b(z)} = j_b^a(z)G. \quad (41)$$

Excluding with the help of the SDE (41) a differentiation over j in SDE (40) and using Bose-symmetry condition

$$\frac{\delta^2 G}{\delta \eta^b_{b'}(y,x)\delta \eta^{b'}_a(x_1,x_1)} = \frac{\delta^2 G}{\delta \eta^b_a(y,x_1)\delta \eta^{b'}_{b'}(x_1,x)}$$

we obtain at $j = 0$ the SDE for generating functional $Z = \log G$:

$$(m_0^2 - \partial_x^2)\frac{\delta Z}{\delta \eta^b_a(y,x)} + \int dx_1 \eta^a_{a'}(x,x_1)\frac{\delta Z}{\delta \eta^b_{a'}(y,x_1)} + \delta^a_b \delta(x-y)$$
$$= -\frac{1}{N}\int dx_1 \alpha(x-x_1)[\frac{\delta^2 Z}{\delta \eta^b_a(y,x_1)\delta \eta^{b'}_{b'}(x_1,x)} + \frac{\delta Z}{\delta \eta^{b'}_{b'}(x_1,x)}\frac{\delta Z}{\delta \eta^b_a(y,x_1)}]. \quad (42)$$

Remind, that $\alpha(x) = g^2 D_c(x)$.

To construct the $1/N$-expansion of this equation, it is convenient to use the Legendre transform of the generating functional $Z[\eta]$. Consider the definition of a propagator with source η

$$\frac{\delta Z}{\delta \eta^b_a(y,x)} = -\Delta^a_b(x,y|\eta) \quad (43)$$

as a functional equation for $\eta = \eta[\Delta]$. By resolving this equation, we can define a new generating functional $\Gamma[\Delta]$ (effective action) from a new functional variable Δ:

$$\Gamma[\Delta] = Z + \int dx_1 dy_1 \Delta^a_b(x_1,y_1)\eta^b_a(y_1,x_1). \quad (44)$$

It is easy to see that

$$\frac{\delta \Gamma}{\delta \Delta^b_a(y,x)} = \eta^a_b(x,y|\Delta). \quad (45)$$

Legendre-transformed SDE (42) can be rewritten as:

$$\frac{\delta \Gamma}{\delta \Delta^b_a(y,x)} = (\Delta^{-1})^a_b(x,y) - (m_0^2 - \partial^2)\delta^a_b(x-y) + \frac{1}{N}\delta^a_b \alpha(x-y)\,\mathrm{tr}\,\Delta(x,y) -$$
$$-\frac{1}{N}\int dx_1 dy_1 \alpha(x-x_1)\frac{\delta \Delta^a_d(x_1,y_1)}{\delta \eta^c_c(x_1,x)}(\Delta^{-1})^d_b(y_1,y). \quad (46)$$

(Here and below $\delta^b_a(x-y) \equiv \delta^b_a \delta(x-y)$.) In this equation $\delta\Delta/\delta\eta$ should be expressed as a functional of Δ. In order to solve this problem, the following connection condition

$$\int dx_1 dy_1 \frac{\delta^2 \Gamma}{\delta \Delta^b_a(y,x)\delta \Delta^d_c(y_1,x_1)}\frac{\delta \Delta^d_c(y_1,x_1)}{\delta \eta^{b'}_{a'}(y',x')} = \delta^a_{b'}(x-y')\delta^{a'}_b(x'-y) \quad (47)$$

is very useful. This connection condition is a consequence of differentiation rule $\delta\eta_b^a(x,y)/\delta\eta_{a'}^{b'}(y',x') = \delta_b^a(x-y')\delta_{b'}^{a'}(x'-y)$ and equations (43) and (45).

SDE (46) tells us a leading approximation for $1/N$-expansion of the generating functional $\Gamma = \Gamma_0 + \Gamma_1 + \cdots$. Since $\Delta_b^a(x,y) = \delta_b^a \Delta(x-y)$ at $\eta = 0$, then $\frac{1}{N}\operatorname{tr}\Delta = \Delta$, and the equation of the leading-order effective action is

$$\frac{\delta\Gamma_0}{\delta\Delta_a^b(y,x)} = (\Delta^{-1})_b^a(x,y) - (m_0^2 - \partial^2)\delta_b^a(x-y) + \frac{1}{N}\delta_b^a \alpha(x-y)\operatorname{tr}\Delta(x,y). \tag{48}$$

Then the NLO equation will be

$$\frac{\delta\Gamma_1}{\delta\Delta_a^b(y,x)} = -\frac{1}{N}\int dx_1 dy_1\, \alpha(x-x_1)\frac{\delta\Delta_{b_1}^a(x_1,y_1)}{\delta\eta_{a_1}^{a_1}(x_1,x)}(\Delta^{-1})_b^{b_1}(y_1,y).$$

One should prove that $\Gamma_1 = O(1/N)$, i.e., $[\delta\Delta/\delta\eta]_0 = O_N(1)$. Differentiating $\delta\Gamma_0/\delta\Delta$ over Δ and using the leading-order connection condition we obtain

$$\left[\frac{\delta\Delta_b^a(x,y)}{\delta\eta_{a'}^{b'}(y',x')}\right]_0 = -\Delta_{b'}^a(x,y')\Delta_b^{a'}(x',y)$$
$$+\frac{1}{N}\int dx_1 dy_1 \Delta_c^a(x,x_1)\Delta_b^c(y_1,y)\alpha(x_1-y_1)\left[\frac{\delta\Delta_d^d(x_1,y_1)}{\delta\eta_{a'}^{b'}(y',x')}\right]_0. \tag{49}$$

Introducing the function

$$Q_b^a\begin{pmatrix} x & x' \\ y & y' \end{pmatrix} \equiv -\left[\frac{\delta\Delta_b^a(x,y)}{\delta\eta_d^d(y',x')}\right]_0\bigg|_{\eta=0}$$

we have the following equation for Q_b^a:

$$Q_b^a\begin{pmatrix} x & x' \\ y & y' \end{pmatrix} = \delta_b^a \Delta(x-y')\Delta(x'-y)$$
$$+\frac{1}{N}\delta_b^a\int dx_1 dy_1 \Delta(x-x_1)\alpha(x_1-y_1)Q_c^c\begin{pmatrix} x_1 & x' \\ y_1 & y' \end{pmatrix}\Delta(y_1-y). \tag{50}$$

Consequently, $Q_b^a = \delta_b^a Q$, where Q is the solution of the equation

$$Q\begin{pmatrix} x & x' \\ y & y' \end{pmatrix} = \Delta(x-y')\Delta(x'-y) + \int dx_1 dy_1 \Delta(x-x_1)\alpha(x_1-y_1)Q\begin{pmatrix} x_1 & x' \\ y_1 & y' \end{pmatrix}\Delta(y_1-y),$$

i.e., $Q = O_N(1)$, Q.E.D.

The phion propagator is

$$\Delta_b^a(x-y) = -\frac{\delta Z_0}{\delta \eta_a^b(y,x)}\Big|_{\eta=0} = \delta_b^a \Delta(x-y)$$

From (48) at $\eta = \delta\Gamma/\delta\Delta = 0$ we have the equation for the leading-order phion propagator

$$\Delta^{-1}(x) = (m_0^2 - \partial_x^2)\delta(x) - \alpha(x)\Delta(x). \qquad (51)$$

The two-particle phion function is

$$Z_{b\,b'}^{a\,a'}\begin{pmatrix} x & x' \\ y & y' \end{pmatrix} = \frac{\delta^2 Z}{\delta\eta_{a'}^{b'}(y',x')\delta\eta_a^b(y,x)}\Big|_{\eta=0} = -\frac{\delta\Delta_b^a(x,y)}{\delta\eta_{a'}^{b'}(y',x')}\Big|_{\eta=0}$$

From (49) we have the leading-order two-particle equation:

$$Z_{b\,b'}^{a\,a'}\begin{pmatrix} x & x' \\ y & y' \end{pmatrix} = \delta_{b'}^a \delta_b^{a'} \Delta(x-y')\Delta(x'-y)$$
$$+\frac{1}{N}\delta_b^a \int dx_1 dy_1 \Delta(x-x_1)\alpha(x_1-y_1)\Delta(y_1-y) Z_{a_1\,b'}^{a_1\,a'}\begin{pmatrix} x_1 & x' \\ y_1 & y' \end{pmatrix}. \qquad (52)$$

The connected part Z_{con} of two-particle function is defined by equation

$$Z_{b\,b'}^{a\,a'}\begin{pmatrix} x & x' \\ y & y' \end{pmatrix} = \delta_{b'}^a \delta_b^{a'} \Delta(x-y')\Delta(x'-y) + \delta_b^a \delta_{b'}^{a'} Z_{con}\begin{pmatrix} x & x' \\ y & y' \end{pmatrix},$$

and the equation for Z_{con} is following:

$$Z_{con}\begin{pmatrix} x & x' \\ y & y' \end{pmatrix} = \frac{g^2}{N}\int dx_1 dy_1 \Delta(x-x_1)\Delta(x'-y_1)D_c(x_1-y_1)\Delta(y_1-y)\Delta(x_1-y')$$
$$+g^2 \int dx_1 dy_1 \Delta(x-x_1) D_c(x_1-y_1)\Delta(y_1-y) Z_{con}\begin{pmatrix} x_1 & x' \\ y_1 & y' \end{pmatrix} \qquad (53)$$

For chion propagator

$$D_{b\,b'}^{a\,a'}(z-z') = \frac{\delta^2 Z}{\delta j_{a'}^{b'}(z')\delta j_a^b(z)}\Big|_{j=\eta=0}$$

we have as a N-component generalization of equation (9) the following equation

$$D_{b\,b'}^{a\,a'}(z-z') = \delta_{b'}^a \delta_b^{a'}[D_c(z-z') + \frac{1}{N}\int dz_1 dz_1' \alpha(z-z_1)\Delta^2(z_1-z_1') D_c(z_1'-z')] +$$

$$+\frac{1}{N}\delta_b^a\delta_{b'}^{a'}\int dz_1 dz_1'\alpha(z-z_1)Z_{con}\begin{pmatrix}z_1 & z_1'\\ z_1 & z_1'\end{pmatrix}D_c(z_1'-z').$$

For the function $D_b^a \equiv D_{b\,a_1}^{a\,a_1}$ we obtain

$$D_b^a(z-z') = \delta_b^a D(z-z'),$$

where D is a solution of equation

$$D(z-z') = D_c(z-z') + \frac{1}{N}\int dz_1 dz_1'\alpha(z-z_1)\Delta^2(z_1-z_1')D_c(z_1'-z')$$
$$+ \int dz_1 dz_1'\alpha(z-z_1)Z_{con}\begin{pmatrix}z_1 & z_1'\\ z_1 & z_1'\end{pmatrix}D_c(z_1'-z'). \quad (54)$$

Since $Z_{con} = O(1/N)$, we have $D = D_c + O(1/N)$, i.e., the chion propagator in the leading-order of $1/N$–expansion is the free propagator.

Three-point function V is defined as a N-component generalization of equation (7):

$$V_{b\,b'}^{a\,a'}(x,y|z) = -\frac{\delta^2 Z}{\delta j_{a'}^{b'}(z)\delta\eta_a^b(y,x)}\Big|_{\eta=0} = \frac{g}{\sqrt{N}}\int dz_1 D_c(z-z_1)Z_{b\,b'}^{a\,a'}\begin{pmatrix}x & z_1\\ y & z_1\end{pmatrix}.$$

For the function $V_b^a \equiv V_{b\,a_1}^{a\,a_1}$ with taking into account the above results we have

$$V_b^a(x,y|z) = \frac{g}{\sqrt{N}}\delta_b^a\int dz_1 D_c(z-z_1)Q\begin{pmatrix}x & z_1\\ y & z_1\end{pmatrix} = \delta_b^a V(x,y|z)$$

where V is a solution of equation

$$V(x,y|z) = \frac{g}{\sqrt{N}}\int dz_1 D_c(z-z_1)\Delta(x-z_1)\Delta(z_1-y)$$
$$+ \int dx_1 dy_1 \Delta(x-x_1)\alpha(x_1-y_1)V(x_1,y_1|z)\Delta(y_1-y).$$

The amputated three-point function (vertex) of the leading order is

$$\Gamma(x,y|z) = \int dx_1 dy_1 dz_1 D_c^{-1}(z-z_1)\Delta^{-1}(x-x_1)V(x_1,y_1|z_1)\Delta^{-1}(y_1-y),$$

and the equation for Γ will be following

$$\Gamma(x,y|z) = \frac{g}{\sqrt{N}}\delta(x-z)\delta(z-y) + \alpha(x-y)\int dx_1 dy_1 \Delta(x-x_1)\Gamma(x_1,y_1|z)\Delta(y_1-y). \tag{55}$$

Going to momentum space

$$\Gamma(x,y|z) = \int \frac{dp_x}{(2\pi)^d}\frac{dp_y}{(2\pi)^d}\frac{dk}{(2\pi)^d} e^{-ixp_x-iyp_y+izk} (2\pi)^d\delta(p_x+p_y-k)\Gamma(p_x|k)$$

we have the leading-order equation for vertex

$$\Gamma(p|k) = \frac{g}{\sqrt{N}} + \int \frac{dq}{(2\pi)^d} \alpha(p-q)\Delta(q)\Gamma(q|k)\Delta(q-k). \tag{56}$$

2.4. n-Particle Approximations

The system of SDEs, generated by functional-differential equation (10), is an infinite set of integral equations for n-particle phion functions $Z_n \equiv \delta^n Z/\delta\eta^n|_{\eta=0}$. The first three SDEs are equations (13), (16) and (17). The nth SDE is the $(n-1)$th derivative of SDE (10) with the source being switched off and includes a set of functions from one-particle function (phion propagator) to the $(n+1)$-particle phion function. In order to obtain a sequence of closed systems of equations, we proceed as follows. We call "the n-particle approximation of the system of SDEs" the system of n SDEs, in which the first $n-1$ equations are exact and the nth SDE is truncated by omitting the $(n+1)$-particle function. It is evident that the sequence of such approximations goes to the exact set of SDEs at $n \to \infty$.

The idea of these approximation scheme is very simple and natural. However, the calculations became more and more complicated at each following stage e.g., the three-particle approximation is a system of three nonlinear equations for the propagator, the two-particle function and the three-particle function.

The one-particle approximation is simply equation (13) without Z_2. This approximation has a trivial solution which is a free propagator. The two-particle approximation is a system of equation (13) and equation (16) without Z_3:

$$(m^2 - \partial_x^2)Z_2^{2P}\begin{pmatrix} x & x' \\ y & y' \end{pmatrix} - \int dx_1 \alpha(x-x_1) Z_2^{2P}\begin{pmatrix} x_1 & x' \\ x_1 & y' \end{pmatrix}\Delta(x-y) +$$
$$= \delta(x-y')\Delta(x'-y). \tag{57}$$

In the framework of the two-particle approximation, Z_2 can be expressed as a functional of Δ. In fact, equation (57) can be written as

$$\int dx_1 dy_1 \, K_{2P}\begin{pmatrix} x & x_1 \\ y & y_1 \end{pmatrix} Z_2^{2P}\begin{pmatrix} x_1 & x' \\ y_1 & y' \end{pmatrix} = \Delta_c(x-y')\Delta(x'-y), \quad (58)$$

where $K_{2P} = K_{12}$ at $\Delta_1 = \Delta_c \equiv (m^2 - \partial^2)^{-1}$ and $\Delta_2 = \Delta$ (see (21)). In correspondence with equations (23)–(26), we obtain

$$Z_2^{2P}\begin{pmatrix} x & x' \\ y & y' \end{pmatrix} = \Delta_c(x-y')\Delta(x'-y) -$$
$$- \int dx_1 dx_2 \, \Delta_c(x-x_1)\Delta(x'-x_2) \, f_{2P}(x_1-x_2)\Delta(x_1-y)\Delta_c(x_2-y'), \quad (59)$$

where f_{2P} is given by equation (26) with $L_{12} \to L_{2P}(x-y) = \Delta_c(x-y)\Delta(y-x)$.

The two-particle approximation as well as the leading term of the mean-field expansion has an incomplete crossing structure of two-particle function. For the two-particle approximation the violation is even more significant because it affects the disconnected part. However, in the same way as for the mean-field expansion, this problem is solved by considering the next approximation. The next three-particle approximation is described by a system of three equations, the first two of which are equations (13) and (16), and the third is equation (17) without Z_4. We can show, that in this approximation the correct cross-symmetric structure is restored for the disconnected part and also for the connected part of the two-particle function. We omit this exercise, referring the reader to work [33].

3. RENORMALIZED MODELS

In this Section, we analyze the behavior in the deep Euclidean region for two models: the complex scalar field model with self-reaction and for the Yukawa model in four dimensions.

3.1. Self-Interacting Scalar Field in Four Dimensions

SDE for theory of complex scalar field ϕ with self-action (12) is

$$(m_0^2 - \partial_x^2)\frac{\delta Z}{\delta\eta(y,x)} + \int dy_1\, \eta(x, y_1)\frac{\delta Z}{\delta\eta(y,y_1)} + \delta(x-y) =$$
$$= \lambda\left[\frac{\delta^2 Z}{\delta\eta(x,x)\delta\eta(y,x)} + \frac{\delta Z}{\delta\eta(x,x)}\frac{\delta Z}{\delta\eta(y,x)}\right]. \quad (60)$$

In this Subsection $x \in E_4$. In correspondence with the construction of Subsection 2.2 the leading-order equation of mean-field expansion will be following

$$(m_0^2 - \partial_x^2)\frac{\delta Z^{(0)}}{\delta\eta(y,x)} + \int dy_1\, \eta(x, y_1)\frac{\delta Z^{(0)}}{\delta\eta(y,y_1)} + \delta(x-y) = \lambda\frac{\delta Z^{(0)}}{\delta\eta(x,x)}\frac{\delta Z^{(0)}}{\delta\eta(y,x)}. \quad (61)$$

Leading-order propagator Δ_0 is given by equation (20) with

$$m^2 = m_0^2 + \lambda\Delta_0(x=0),$$

and the leading-order two-particle function is given by equation (28) with

$$f_0(p) = \frac{\lambda}{1 + \lambda L_0(p)},$$

where

$$L_0(p) = \int \frac{d^4q}{(2\pi)^4}\, \Delta_0(p+q)\Delta_0(q)$$

is the single scalar loop.

All above formulae for the propagator Δ_0 and the amplitude f_0 contain divergent integrals and should be renormalized. In correspondence with the standard recipe we introduce the renormalized Lagrangian

$$\mathcal{L} = -\partial_\mu\phi^*\partial_\mu\phi - m^2\phi^*\phi - \frac{\lambda}{2}(\phi^*\phi)^2, \quad (62)$$

where ϕ, m and λ are the renormalized field, mass and coupling, and add the counter-terms

$$\Delta\mathcal{L} = -(z_\phi - 1)\partial_\mu\phi^*\partial_\mu\phi - \delta m^2\phi^*\phi - (z_\lambda - 1)\frac{\lambda}{2}(\phi^*\phi)^2, \quad (63)$$

which absorb the divergences.

Full (or, bare) Lagrangian $\mathcal{L}_b = \mathcal{L} + \Delta\mathcal{L}$ can be written as

$$\mathcal{L}_b = -\partial_\mu \phi_b^* \partial_\mu \phi_b - m_b^2 \phi_b^* \phi_b - \frac{\lambda_b}{2}(\phi_b^* \phi_b)^2, \tag{64}$$

where

$$\phi_b = \sqrt{z_\phi}\phi, \quad m_b^2 = \frac{m^2 + \delta m^2}{z_\phi}; \quad \lambda_b = \frac{\lambda z_\lambda}{z_\phi^2}. \tag{65}$$

Then all above calculations are reproduced with bare Lagrangian (64), i.e., with the replacement $m_0^2 \to m_b^2$, $\lambda \to \lambda_b$, $\Delta_0 \to \Delta_b$ etc., and the normalization conditions are imposed on the renormalized propagator Δ_r and amplitude f_r. For the easement of the following calculations we choose the normalization point at zero momenta.

The normalization conditions for the propagator $\Delta_r(p^2) = z_\phi^{-1}\Delta_b(p^2)$ are

$$\Delta_r^{-1}(0) = m^2, \tag{66}$$

$$\frac{d}{dp^2}\Delta_r^{-1}\Big|_{p^2=0} = 1. \tag{67}$$

These conditions define the mass-renormalization counter-term δm^2 and the field-renormalization constant z_ϕ. It is easy to see that $z_\phi = 1$ in the case, and, consequently, $\Delta^{-1} = \Delta_b^{-1} - m^2 + p^2$, $Z_2^{(0)} = Z_{2b}^{(0)}$. Thus, the amplitude $f_r = f_b$ is

$$f_r(p) = \frac{\lambda z_\lambda}{1 + \lambda z_\lambda L_0(p)}. \tag{68}$$

The normalization condition for the amplitude is following

$$f_r(0) = \lambda. \tag{69}$$

This condition defines the coupling-renormalization constant z_λ and the renormalized amplitude

$$f_r(p) = \frac{\lambda}{1 + \lambda L_r(p)}, \tag{70}$$

where

$$L_r(p) = L_0(p) - L_0(0) = -\frac{1}{16\pi^2}\int_0^1 dz \log(1 + z(1-z)\frac{p^2}{m^2}) \tag{71}$$

is the renormalized loop.

As it follows from equations (70) and (71) the renormalized amplitude f_r possesses a non-physical singularity (Landau pole) in the point $p^2 = M_L^2$, where M_L^2 is a solution of the equation

$$1 + \lambda L_r(M_L^2) = 0.$$

This equation has a solution at any positive λ. At $p^2 \to \infty$

$$L_r \simeq -\frac{1}{16\pi^2} \log \frac{p^2}{m^2}$$

and $M_L \cong m \exp\{\frac{8\pi^2}{\lambda}\}$. As it was yet noted in Introduction, the same Landau pole arises at the calculations of renormalized amplitude by other methods: in the frameworks of $1/N$-expansion and at the renormalization-group summation.

The situation changes when considering a two-particle approximation of a Subsection 2.4.

In correspondence with the general construction, the two-particle approximation for the complex scalar field with self-action (12) is described by the following system of non-linear equations

$$\begin{cases} (m^2 - \partial_x^2)\Delta(x-y) = \delta(x-y) - \lambda Z_2\begin{pmatrix} x & x \\ y & x \end{pmatrix} \\ (m^2 - \partial_x^2)Z_2\begin{pmatrix} x & x' \\ y & y' \end{pmatrix} + \lambda Z_2\begin{pmatrix} x & x' \\ x & y' \end{pmatrix}\Delta(x-y) - \delta(x-y')\Delta(x'-y) \end{cases} \quad (72)$$

As explained above, the two-particle function Z_2 can be considered as a functional of Δ (see equation (59))

$$Z_2\begin{pmatrix} x & x' \\ y & y' \end{pmatrix} = \Delta_c(x-y')\Delta(x'-y) -$$

$$- \int dx_1 dx_2 \Delta_c(x-x_1)\Delta(x'-x_2)f(x_1-x_2)\Delta(x_1-y)\Delta_c(x_2-y'). \quad (73)$$

Here function $f(p)$ in the momentum space is a solution of equation

$$\frac{1}{f(p)} = \frac{1}{\lambda} + \int \frac{d^4q}{(2\pi)^4}\Delta_c(p+q)\Delta(q), \quad (74)$$

where

$$\Delta_c(p) = \frac{1}{m^2 + p^2}. \quad (75)$$

First equation of system (72) can be rewritten as

$$\Delta^{-1}(x-y) = (m^2 - \partial_x^2)\delta(x-y) + \Sigma(x-y), \tag{76}$$

where mass operator Σ is

$$\Sigma(x-y) = \lambda \int dx_1 Z_2\left(\begin{array}{cc} x & x \\ x_1 & x \end{array}\right) \Delta^{-1}(x_1 - y). \tag{77}$$

Taking into account equations (73) and (74) we obtain for Σ in the momentum space the following expression

$$\Sigma(p) = \int \frac{dq}{(2\pi)^4} \Delta_c(p-q) f(q). \tag{78}$$

If we approximate propagator Δ by free propagator Δ_c:

$$\Delta(p) \approx \Delta_c(p). \tag{79}$$

then we obtain as a solution of equation (74) the mean-field amplitude f_0. So the mean-field approximation is contained in the two-particle approximation as a first iteration.

The renormalization of the system of equations (72) can be done in correspondence with the general recipe by introducing counter-term Lagrangian (63) and bare Lagrangian (64). In terms of the bare quantities the bare propagator is

$$\Delta_b^{-1}(p^2) = m_b^2 + \lambda_b \Delta_b(x=0) + p^2 + \Sigma_b(p^2), \tag{80}$$

where

$$\Sigma_b(p^2) = \int \frac{d^4q}{(2\pi)^4} \Delta_c(p-q) f_b(q) \tag{81}$$

is the bare mass operator.

The normalization conditions (66) and (67) for the renormalized propagator $\Delta = z_\phi^{-1}\Delta_b$ define the mass-renormalization counter-term δm^2 and field-renormalization constant

$$z_\phi = (1 + \Sigma_b'(0))^{-1}. \tag{82}$$

The equation for renormalized propagator is

$$(m^2 + p^2)\Delta(p^2) = 1 - \Delta(p^2)\Sigma_r(p^2), \tag{83}$$

where
$$\Sigma_r(p^2) = z_\phi[\Sigma_b(p^2) - \Sigma_b(0) - p^2 \Sigma_b'(0)] \qquad (84)$$
is the renormalized mass operator.

The equation for f_b is
$$\frac{1}{f_b(p)} = \frac{1}{\lambda_b} + L_b(p), \qquad (85)$$
where
$$L_b(p) = \int \frac{d^4q}{(2\pi)^4} \Delta_c(p+q) \Delta_b(q) = z_\phi \int \frac{d^4q}{(2\pi)^4} \Delta_c(p+q) \Delta(q) \qquad (86)$$
is the bare loop operator.

The renormalized coupling λ is defined as above by condition (69). This normalization condition together with equation (85) define the coupling-renormalization constant z_λ and the renormalized equation for f_r:
$$\frac{1}{f_r(p^2)} = \frac{1}{\lambda} + L_r(p^2), \qquad (87)$$
where
$$L_r(p^2) = L_b(p^2) - L_b(0) \qquad (88)$$
is the renormalized loop operator.

Equations (83) and (87) are the system of non-linear integral equations for the propagator and the amplitude. This system can be solved by the expansion in the vicinity of the point $p = 0$. Such solution, however, is not interesting because it is a some part of the usual perturbation theory over the coupling λ.

Much more interesting problem is to look for the asymptotic behavior of the solution at large momenta. In the large-momenta region an essential technical simplification is possible, namely, one can replace in integrals (81) and (86) the function Δ_c (see (75)) by massless function $1/p^2$:
$$\int \frac{d^4q}{(2\pi)^4} \Phi(q^2) \Delta_c(p-q) \Longrightarrow \int \frac{d^4q}{(2\pi)^4} \frac{\Phi(q^2)}{(p-q)^2}. \qquad (89)$$
Then it is possible to use the well-known formula
$$\int \frac{d^4q}{(2\pi)^4} \frac{\Phi(q^2)}{(p-q)^2} = \frac{1}{16\pi^2} \left[\frac{1}{p^2} \int_0^{p^2} \Phi(q^2) q^2 \, dq^2 + \int_{p^2}^{\infty} \Phi(q^2) \, dq^2 \right]. \qquad (90)$$

This massless-integration approximation is quite usual in investigations in the deep-Euclidean region, though rigorous arguments for its validity can be done for the asymptotically-free models only [13]. In the general case, this approximation should be considered as a plausible conjecture, which needs further investigations. Formula (90) highly enables the calculations and, as a major point, permits us to go from integral equations to differential ones (see below).

Further essential simplification of the system of equations consists in replacing of equation (83) by approximate relation

$$(m^2 + p^2)\Delta(p^2) \approx 1 - \Delta_c(p^2)\Sigma_r(p^2). \tag{91}$$

This approximation enables "to unleash" the system, keeping at the time the non-linearity, i.e., to obtain the non-linear integral equation for f_r, which does not contain Δ. This approximation, of course, is less substantiated and should be considered as an iteration of the initial equations.

Approximation (91) defines the closed equation for f_r, which will be a main object of the following consideration. For the future convenience we introduce new quantities w, $\bar{\lambda}$ and dimensionless variable t as[6]

$$w = \frac{z_\phi}{16\pi^2} f_r, \quad \bar{\lambda} = \frac{z_\phi}{16\pi^2} \lambda, \quad t = \frac{p^2}{m^2}. \tag{92}$$

The normalization condition for $w(t)$ is

$$w(0) = \bar{\lambda}. \tag{93}$$

Simple calculation lead to the equation for $w(t)$:

$$\frac{1}{w} = \frac{1}{\bar{\lambda}} + \left(\frac{\bar{\lambda}}{2} - 1\right)\log(1+t) + (1-\bar{\lambda})\left(1 - \frac{1}{t}\log(1+t)\right) + \int_0^t dt_1 \mathcal{K}(t|t_1) w(t_1), \tag{94}$$

where the kernel $\mathcal{K}$ is

$$\mathcal{K}(t|t_1) = \frac{t_1}{t} - 1 + \frac{1}{t}\log\frac{1+t}{1+t_1} + t_1 \log\frac{t(1+t_1)}{t_1(1+t)}. \tag{95}$$

---

[6] As has been repeatedly noted (see, for example, [26], footnote 3), the $16\pi^2$ factor in the definition of $\bar{\lambda}$ corresponds to the "natural" normalization of the coupling constant in the theory of a scalar field with quartic interaction. With this normalization, the critical coupling constant (see below) becomes a value of the order of unity.

Integral equation (94) is the non-linear Volterra equation and can be reduced to non-linear differential equation of fourth order

$$\frac{d^2}{dt^2}\left[t(t+1)^2 \frac{d^2}{dt^2}\left(\frac{t}{w}\right)\right] = \bar{\lambda} - 2 - w. \tag{96}$$

Initial conditions in the point $t = 0$ for this equations can be easily derived from integral equation (94).

As it easy to see, differential equation (96) at $\bar{\lambda} \neq 2$ has the exact solution

$$w_{exact} = \bar{\lambda} - 2. \tag{97}$$

This solution is the leading term of asymptotic expansion at $t \to \infty$ of the solution of integral equation (94). Indeed, a simple calculation demonstrates that after the substitution of (97) into (94) the leading increasing logarithmic term is cancelled. With substisution

$$w = \bar{\lambda} - 2 + \varphi(t), \tag{98}$$

where $\varphi(t) \to 0$ at $t \to \infty$, we have for the function $\varphi(t)$ at large t the linear differential equation

$$\frac{d^2}{dt^2}\left[t^3 \frac{d^2}{dt^2}(t\varphi)\right] = (\bar{\lambda} - 2)^2 \varphi. \tag{99}$$

The solutions of this Euler-type equation are t^{c-1}, where

$$c^2(c^2 - 1) = (\bar{\lambda} - 2)^2. \tag{100}$$

This characteristic equation has two real and two imaginary roots. The first real root is greater than 1, and corresponds to an increasing solution. The second real root is less than -1, and corresponds to a fast-decreasing solution. The imaginary roots correspond to decreasing as t^{-1} and oscillating solutions. Hence, just these roots define the next-to-the-leading asymptotic term of the solution of integral equation (94).

For further calculation it is convenient to go to the variable

$$x = \log(1 + t) \tag{101}$$

The next-to-the-leading asymptotic term in terms of this variable is

$$\varphi = e^{-x}(A \cos \omega x - B \sin \omega x), \tag{102}$$

where
$$\omega = \sqrt{\sqrt{(\bar{\lambda}-2)^2 + \frac{1}{4}} - \frac{1}{2}}. \qquad (103)$$

Substituting (98) and (102) into integral equation (94) we obtain at $t \to \infty$ the relation
$$\frac{1}{\bar{\lambda}-2} = \frac{1}{\bar{\lambda}} - 1 - \varsigma(\omega)A + \omega\tau(\omega)B, \qquad (104)$$
where
$$\varsigma(\omega) = \sum_{k=1}^{\infty} \frac{1}{(k+1)(k^2+\omega^2)}, \quad \tau(\omega) = \sum_{k=1}^{\infty} \frac{1}{k(k+1)(k^2+\omega^2)}. \qquad (105)$$

Due to normalization condition (93) we have $A = 2$, and equation (104) defines the quantity $B(\bar{\lambda})$.

At $t \to \infty$ we have $y \to \bar{\lambda} - 2$. At $\bar{\lambda} < 2$ such behavior contradicts to continuity. If the function y is continuous, then at $\bar{\lambda} < 2$ $y(t_0) = 0$ in some point $t_0 \in (0; +\infty)$ (since y changes the sign). But in the case integral equation (94) is not fulfilled. Consequently the function y is singular in some point of pre-asymptotic region (Landau pole, or something similar), and at $\bar{\lambda} < 2$ this model is inconsistent beyond the region of small momenta.

Hence, the region of self-consistence of this model is $\bar{\lambda} > 2$. Using equation (82) is easy to prove that $z_\phi = 2$ at $\bar{\lambda} = 2$. Then
$$\lambda_{cr} = 16\pi^2. \qquad (106)$$

Therefore, at $\lambda > \lambda_{cr}$ the asymptotic behavior of amplitude at large t are given as follows
$$f_r \sim (\lambda - \lambda_{cr}) + \varphi(t), \qquad (107)$$
where $\varphi(t) = O(1/t)$ is given by equations (102) – (105).

The most interesting result of this Subsection is, of course, the existence of the critical point λ_{cr}, which divides the whole set of coupling values in two region: the weak-coupling region of inconsistency and the strong-coupling region of non-trivial self-consistent behavior.[7] This division of the coupling values,

---

[7]Note that a singular behavior on parameters for the theory of four-dimensional scalar field was marked also by Halpern and Huang [29] which have found a class of non-trivial interactions of scalar field in four dimensions. As pointed by Morris [50], such interactions correspond to singular effective potential. The connection of this fact with our result is, however, unclear due to very different methods of investigation.

which detected for two-particle approximation, cannot be indicated by the expansions tightly connected with perturbation theory (e.g., the renormalization-group summation and the $1/N$-expansion).

3.2. Yukawa Model

In this Subsection we consider the theory of a Dirac fermion field ψ interacting with a pseudoscalar boson field χ in a four-dimensional (1+3) space ($x \in M_4$) with the Lagrangian

$$\mathcal{L} = \bar{\psi}(i\hat{\partial} - m_\psi)\psi - \frac{1}{2}\chi(\mu^2 + \partial^2)\chi + g_y\bar{\psi}\Gamma\psi\chi. \tag{108}$$

Here $\Gamma = i\gamma^5$ and $\hat{\partial} = \gamma^\mu\partial_\mu$; γ^μ are Dirac matrices and g_y is Yukawa coupling. We suppose $m_\psi > 0$. A case $m_\psi = 0$ needs a special consideration and is not discussed here.

The renormalizability requires to supplement Lagrangian (108) with an additional term $\lambda\chi^4$ which corresponds to the self-interaction of the boson field. This term ensures a renormalization of the boson-boson scattering amplitude. In this work we are not concerned with this amplitude, and the renormalization of approximations considered below does not require including the corresponding counter-term. For this reason we do not include the quartic interaction into consideration. Therefore, we shall consider the restricted Yukawa model neglecting the self-interaction of the boson field, and the obtained results should be treated as the first step to the study of the asymptotic behavior in a realistic model of the boson-fermion interaction. In other words, we shall consider the case $\lambda = 0$ only.

The generating functional can be written as a functional integral

$$G = \int D(\psi, \bar{\psi}, \chi) \exp i\left[\int dx\, \mathcal{L} - \int dxdy\, \bar{\psi}(y)\eta(y,x)\psi(x) + \int dx\, j(x)\chi(x)\right], \tag{109}$$

where $j(x)$ is a single boson source and $\eta(x, y)$ is a bilocal fermion source. Here and everywhere further in this Subsection we will, as usual, omit the spinor indices.

The translational invariance of the functional integration measure leads to the functional-differential SDEs for generating functional G. The derivation of these equations is completely similar to the derivation of analogous SDEs in the Subsection 2.1 (with taking into account the Grassmanian nature of field ψ). In terms of the logarithm $Z = \frac{1}{i}\log G$ these equations are:

$$\delta(x-y) + (i\hat{\partial}_x - m_\psi)i\frac{\delta Z}{\delta\eta(y,x)} + g_y\Gamma\left[\frac{\delta^2 Z}{\delta\eta(y,x)\delta j(x)} + i\frac{\delta Z}{\delta\eta(y,x)}\frac{\delta Z}{\delta j(x)}\right] =$$
$$= \int dx_1 \eta(x,x_1) i\frac{\delta Z}{\delta\eta(y,x_1)}, \quad (110)$$

$$\frac{\delta Z}{\delta j(x)} = \int dx_1 \left[D_c(x-x_1)j(x_1) - g_y D_c(x-x_1)\,\mathrm{tr}\left(\Gamma\frac{\delta Z}{\delta\eta(x_1,x_1)}\right)\right]. \quad (111)$$

Here $D_c = (\mu^2 + \partial^2)^{-1}$. We define also the fermion propagator

$$S(x-y) = i\frac{\delta Z}{\delta\eta(y,x)}\bigg|_{\eta=j=0}, \quad (112)$$

the boson propagator

$$D(x-y) = \frac{\delta^2 Z}{\delta j(y)\delta j(x)}\bigg|_{\eta=j=0}, \quad (113)$$

the two-particle (four-point) fermion function

$$Z_2\begin{pmatrix} x & x' \\ y & y' \end{pmatrix} = i\frac{\delta^2 Z}{\delta\eta(y',x')\delta\eta(y,x)}\bigg|_{\eta=j=0} \quad (114)$$

and the three-point function

$$V(z|x,y) = i\frac{\delta^2 Z}{\delta\eta(y,x)\delta j(z)}\bigg|_{\eta=j=0}. \quad (115)$$

Differentiations of SDE (111) over η and j give us the SDE for the three-point function

$$V(z|x,y) = -g_y \int dz_1 D_c(z-z_1)\Gamma\, Z_2\begin{pmatrix} x & z_1 \\ y & z_1 \end{pmatrix} \quad (116)$$

and the SDE for the boson propagator

$$D(x-y) = D_c(x-y) + ig_y \int dy_1\,\mathrm{tr}\left[\Gamma V(x|y_1,y_1)\right] D_c(y_1-y). \quad (117)$$

Excluding with the help of SDE (111) a differentiation over j in SDE (110), we obtain at $j = 0$ the SDE for the generating functional:

$$\delta(x-y) + (i\hat{\partial}_x - m_\psi)i\frac{\delta Z}{\delta\eta(y,x)} = \int dx_1 \Big\{ i\eta(x,x_1)\frac{\delta Z}{\delta\eta(y,x_1)} +$$
$$+\alpha_y(x-x_1)\Gamma\Big[i\frac{\delta Z}{\delta\eta(y,x)}\operatorname{tr}\Big(\Gamma\frac{\delta Z}{\delta\eta(x_1,x_1)}\Big) + \frac{\delta}{\delta\eta(y,x)}\operatorname{tr}\Big(\Gamma\frac{\delta Z}{\delta\eta(x_1,x_1)}\Big)\Big]\Big\} \quad (118)$$

which contains only the derivatives over the bilocal source η. Here $\alpha_y = g_y^2 D_c$. Switching off the source η in (118), we have the SDE for the fermion propagator

$$(m_\psi - i\hat{\partial}_x) S(x-y) = \delta(x-y) + i\int dx_1 \alpha_y(x-x_1)\Gamma Z_2 \begin{pmatrix} x & x_1 \\ y & x_1 \end{pmatrix} \Gamma. \quad (119)$$

A differentiation of (118) over η gives us (with the source being switched off) the SDE for the two-particle fermion function

$$(m_\psi - i\hat{\partial}_x) Z_2 \begin{pmatrix} x & x' \\ y & y' \end{pmatrix} + \delta(x-y')S(x'-y) =$$
$$= i\int dx_1 \Big\{ (\Gamma S(x-y))\alpha_y(x-x_1)Z_2 \begin{pmatrix} x_1 & x' \\ x_1 & y' \end{pmatrix} \Gamma + \alpha_y(x-x_1)\Gamma Z_3 \begin{pmatrix} x & x_1 & x' \\ y & x_1 & y' \end{pmatrix} \Gamma \Big\} \quad (120)$$

Here $Z_3 = i\frac{\delta^3 Z}{\delta\eta^3}\big|_{\eta=0}$ is the three-particle (six-point) fermion function. The derivation of equation (120) implies that $\operatorname{tr}(\gamma^5 S) = 0$, i.e., we suppose parity conservation.

To construct the mean-field expansion, we consider as a leading approximation for equation (118) the equation

$$\delta(x-y) + (i\hat{\partial}_x - m_\psi)i\frac{\delta Z^{MF}}{\delta\eta(y,x)} =$$
$$i\int dx_1 \Big\{\eta(x,x_1)\frac{\delta Z^{MF}}{\delta\eta(y,x_1)} + \alpha_y(x-x_1)\Gamma\frac{\delta Z^{MF}}{\delta\eta(y,x)}\operatorname{tr}\Big[\Gamma\frac{\delta Z^{MF}}{\delta\eta(x_1,x_1)}\Big]\Big\} (121)$$

The mean-field fermion propagator is

$$S = S^c, \quad (122)$$

where $S^c = (m_\psi - i\hat{\partial})^{-1}$.

Equation (121) gives the equation for the two-particle function

$$Z_2 \begin{pmatrix} x & y \\ x' & y' \end{pmatrix} + S^c(x-y')S^c(x'-y) =$$
$$= i\int dx_1 dx_2 (S^c(x-x_1)\Gamma S^c(x_1-y))\alpha_y(x_1-x_2)Z_2 \begin{pmatrix} x_2 & x_2 \\ x' & y' \end{pmatrix}\Gamma \quad (123)$$

whose solution is

$$Z_2\begin{pmatrix} x & y \\ x' & y' \end{pmatrix} = -S^c(x-y')S^c(x'-y) +$$
$$+ \int dx_1 dx_2 (S^c(x-x_1)\Gamma S^c(x_1-y)) f_{MF}(x_1-x_2)(S^c(x'-x_2)\Gamma S^c(x_2-y')), \quad (124)$$

where in the momentum space

$$\frac{1}{f_{MF}(p^2)} = \frac{i}{\alpha_y(p^2)} + L^c(p^2), \quad (125)$$

and

$$L^c(p^2) = \int \frac{d^4q}{(2\pi)^4} \, \text{tr}\,[S^c(p+q)\Gamma S^c(q)\Gamma] \quad (126)$$

is the single fermion loop.

Taking into account SDEs (116) and (117), we obtain the mean-field boson propagator

$$D_{MF}(p^2) = \frac{i}{g_y^2} f_{MF}(p^2). \quad (127)$$

A renormalization of the mean-field approximation can be made by introducing counter-terms in the Lagrangian in correspondence with the standard recipe of preceding Subsection. The normalization conditions for the propagator $D_r(p^2)$ are

$$D_r^{-1}(0) = \mu^2, \quad \frac{d}{dp^2} D_r^{-1}\Big|_{p^2=0} = 1. \quad (128)$$

These conditions define a mass-renormalization counter-term and a field-renormalization constant. Then the renormalized boson propagator is defined as

$$D_r^{-1}(p^2) = \mu^2 - p^2 - ig_y^2 L_r^c(p^2) \quad (129)$$

where

$$L_r^c(p^2) = L^c(p^2) - L^c(0) - p^2 (L^c)'(0) = \frac{ip^2}{8\pi^2} \int_0^1 dz \log[1 - z(1-z)\frac{p^2}{m^2}]. \quad (130)$$

is the renormalized fermion loop. As it follows from equation (129), the renormalized boson propagator D_r possesses a non-physical singularity (Landau pole) in the Euclidean region $p^2 < 0$ at the point $p_e^2 \equiv -p^2 = M_L^2$, where M_L^2 is a solution of the equation

$$\mu^2 + M_L^2 - ig_y^2 L_r^c(-M_L^2) = 0.$$

This equation has a solution at any positive g_y^2. As was yet noted in the introduction, the same Landau pole arises in the calculations of the renormalized amplitude by other methods: in the frameworks of $1/N$-expansion and renormalization-group summation.

The two-particle approximation for the Yukawa model is the system of equation (119) and equation (120) without Z_3:

$$\begin{cases} (m_\psi - i\hat{\partial}_x) S(x-y) = \delta(x-y) + i \int dx_1 \alpha_y(x - x_1) \Gamma Z_2 \begin{pmatrix} x & x_1 \\ y & x_1 \end{pmatrix} \\ (m_\psi - i\hat{\partial}_x) Z_2 \begin{pmatrix} x & y \\ x' & y' \end{pmatrix} + \delta(x-y') S(x'-y) = i \int dx_1 (\Gamma S(x-y)) \alpha_y(x-x_1) Z_2 \begin{pmatrix} x_1 & x_1 \\ x' & y' \end{pmatrix} \Gamma \end{cases} \tag{131}$$

which includes S and two-particle function Z_2. In system (131), the two-particle function Z_2 can be considered as a functional of S:

$$Z_2 \begin{pmatrix} x & y \\ x' & y' \end{pmatrix} = -S^c(x-y') S(x'-y)$$

$$+ \int dx_1 dx_2 \, (S^c(x-x_1) \Gamma S(x_1-y)) f_y(x_1-x_2)(S(x'-x_2) \Gamma S^c(x_2-y')), \tag{132}$$

where

$$f_y(x-y) = -i\alpha_y(x-y) + i \int dx_1 dx_2 \, \alpha_y(x-x_1) L_F(x_1-x_2) f_y(x_2-y) \tag{133}$$

and $L_F(x) = \mathrm{tr}\,[S^c(x)\,\Gamma\,S(-x)\,\Gamma]$ is the fermion loop operator. In momentum space:

$$\frac{1}{f_y(p^2)} = i\alpha_y^{-1}(p^2) + L_F(p^2), \tag{134}$$

$$L_F(p^2) = \int \frac{d^4q}{(2\pi)^4} \, \mathrm{tr}\,[S^c(p+q)\,\Gamma\,S(q)\,\Gamma]. \tag{135}$$

Taking into account equations (132)–(135), (116) and (117), we obtain for the boson propagator D the following equation in momentum space:

$$D^{-1}(p^2) = \mu^2 - p^2 - ig_y^2 L_F(p^2). \tag{136}$$

From equation (119) and equations (132)–(136) we have the equation for the fermion propagator

$$S^{-1}(p) = m_\psi - \hat{p} + ig_y^2 M(p), \tag{137}$$

where
$$M(p) = \int \frac{d^4q}{(2\pi)^4} \Gamma S^c(p-q) \Gamma D(q). \qquad (138)$$

The system of equations (135)–(138) is the system of unrenormalized SDEs in the two-particle approximation.

The renormalization of equations (136) and (137) can be performed in correspondence with the general recipe by introducing the counter-term Lagrangian. Normalization conditions (128) for the propagator $D_r(p^2)$ and for the fermion propagator

$$S_r^{-1}(p=0) = m_\psi, \quad \left.\frac{\partial S_r^{-1}(p)}{\partial \hat{p}}\right|_{p=0} = -1 \qquad (139)$$

define the counter-terms, and the system of renormalized equations will be

$$\begin{cases} S_r^{-1}(p) = m_\psi - \hat{p} + ig_y^2 M_r(p) \\ D_r^{-1}(p^2) = \mu^2 - p^2 - ig_y^2 L_{rF}(p^2) \end{cases} \qquad (140)$$

where

$$M_r(p) = M(p) - M(0) - \hat{p}\left.\frac{\partial M}{\partial \hat{p}}\right|_{p=0}, \qquad (141)$$

$$L_{rF}(p^2) = L_F(p^2) - L_F(0) - p^2 L'_F(0). \qquad (142)$$

In equations (140) and below m_ψ and μ are the renormalized masses, g_y is the renormalized coupling.

As with the scalar field model discussed in the previous Subsection, an iteration of equation for D_r in (140) with $S^{(0)} = S^c$ leads to the mean-field propagator (129). So the mean-field approximation is contained in the two-particle approximation as the first iteration.

System of equations (140) is the system of nonlinear integral equations for the propagators. The most interesting problem is to look for the asymptotic behavior of the solution of system (140) at large Euclidean momenta. As in the previous section, an essential technical simplification is possible in the large-momenta region. Namely, one can replace in integrals (135) and (138) the function S_c by a massless function $-1/\hat{p}$:

$$\int \frac{d^4q}{(2\pi)^4} F(q) S_c(p-q) \implies -\int \frac{d^4q}{(2\pi)^4} \frac{F(q)}{\hat{p}-\hat{q}}. \qquad (143)$$

Then it is possible (after going to Eucliden momentum space) to use massless-integration formula (90). Equations (140) in this approximation in the Euclidean region are

$$\begin{cases} a(p_e^2) = 1 - \frac{g_y^2}{32\pi^2 (p_e^2)^2} \int_0^{p_e^2} dq_e^2 (p_e^2 - q_e^2)^2 D_r(q_e^2) \\ D_r^{-1}(p_e^2) = \mu^2 + p_e^2 - \frac{g_y^2}{8\pi^2 p_e^2} \int_0^{p_e^2} dq_e^2 (p_e^2 - q_e^2)^2 \frac{a(q_e^2)}{m_\psi^2 + q_e^2 a^2} \end{cases} \quad (144)$$

Here $p_e \in E_4$, and $S_r^{-1} = m_\psi - a(p_e^2)\hat{p}$ in the massless-integration approximation.

Introducing the dimensionless quantities $t = \frac{p_e^2}{m_\psi^2}$, $h(t) = \frac{1}{p_e^2 D_r}$, system (144) can be written as follows

$$\begin{cases} a(t) = 1 - \frac{g_y^2}{32\pi^2} \int_0^t (1 - \frac{t_1}{t})^2 \frac{dt_1}{t_1 h(t_1)} \\ h(t) = 1 + \frac{\mu^2}{m_\psi^2 t} - \frac{g_y^2}{8\pi^2} \int_0^t (1 - \frac{t_1}{t})^2 \frac{dt_1\, a(t_1)}{1 + t_1 a^2(t_1)} \end{cases} \quad (145)$$

This system (multiplied to t^2) after threefold differentiations is reduced to the system of differential equations

$$\begin{cases} \frac{d^3}{dt^3}(t^2 a) = -\frac{g_y^2}{16\pi^2} \frac{1}{th} \\ \frac{d^3}{dt^3}(t^2 h) = -\frac{g_y^2}{4\pi^2} \frac{a}{1 + a^2 t} \end{cases} \quad (146)$$

System of integral equations (145) give us boundary conditions for system (146) at the point $t = 0$.

At large t, system (146) has the asymptotic solution

$$a \simeq \mathcal{A} \log^{1/5} t, \quad h \simeq \mathcal{B} \log^{4/5} t, \quad \mathcal{AB} = -\frac{5g_y^2}{32\pi^2}. \quad (147)$$

System of differential equations (146) do not fix the signs of coefficients $\mathcal{A}$ and $\mathcal{B}$. These signs have the principal meaning and define the physical situation, described by two-particle approximation. If $\mathcal{A} > 0$ ($\mathcal{B} < 0$), then function $h(t)$ applies to zero at some point. This case corresponds to the presence of Landau-type singularity in the boson propagator, i.e., the situation is similar to the above physically unsatisfactory MF approximation. Oppositely, if $\mathcal{A} < 0$ ($\mathcal{B} > 0$), then the situation corresponds to self-consistent asymptotic behavior of the boson propagator. In this case, function $a(t)$ has a zero at some point, but since $1 + ta^2 > 0$ at any $t > 0$, it does not lead to a Landau pole in the propagator for massive fermions.

A study of this system under very general assumptions shows that with taking into account the boundary conditions, implied by integral equations (145), the second possibility is being realized [8], i.e., $\mathcal{A} < 0$ ($\mathcal{B} > 0$). At $t \gg 1$

$$\mathcal{A} \approx -0.77 \frac{g_y^2}{16\pi^2}, \quad \mathcal{B} \approx 3.24, \tag{148}$$

and

$$D_r \simeq \frac{0.308}{p_e^2 \log^{4/5} \frac{p_e^2}{m^2}}. \tag{149}$$

We see that for the Yukawa model instead of physically unsatisfactory behavior with the Landau pole in the Euclidean region, which occurs for the mean-field approximation, the boson propagator in the two-particle approximation has self-consistent asymptotic behavior, which is similar to the asymptotically free behavior.

4. SUPER-RENORMALIZED MODEL: SCALAR YUKAWA MODEL IN FOUR DIMENSIONS

In this Section we consider a vector-matrix model of the complex scalar field ϕ_a (phion) and real scalar mass-less field χ_{ab} (chion) with interaction $g\phi_a^* \phi_b \chi_{ab}$ in four dimensions ($x \in E_4$; $a, b = 1, \cdots, N$.) The Lagrangian of the model is given by equation (39) at $\mu_0 = 0$.

We obtain a solution of the equation for the phion propagator in the leading order of the $1/N$ – expansion, and also investigate the asymptotic behavior of the vertex for zero transfer momentum. The solution of the phion-propagator equation show a change of the asymptotic behavior in the deep Euclidean region in a vicinity of a certain critical value of the coupling constant. For small coupling the propagator behaves asymptotically as free. In the strong-coupling region the propagator in the deep Euclidean region tend to some constant limit. At the critical value of coupling that separates the weak and strong coupling regions, the asymptotic behavior of the propagator is $\sim 1/p$, i.e., it is a medium among the free behavior and the constant–type behavior in strong–coupling region.

---

[8] A detailed presentation of this study is contained in the author's work [32].

The similar change of asymptotic behavior for this model was found also with solution of the system of Schwinger–Dyson equations in two-particle approximation [33] and in ladder approximation [34]. (The latter is, in fact, the leading approximation of $1/N$–expansion at $N = 1$.)

We also discuss in this Section an analogy between the phase transition in the model under consideration and the re-arrangement of the physical vacuum in the supercritical external field.

The leading-order equations for phion propagator Δ and vertex Γ in the momentum space equations are (see Subsection 2.3)

$$\Delta^{-1}(p) = m^2 + p^2 - g^2 \int \frac{d^4q}{(2\pi)^4} D_c(p-q)\Delta(q), \qquad (150)$$

and

$$\Gamma(p|k) = \frac{g}{\sqrt{N}} + g^2 \int \frac{d^4q}{(2\pi)^4} D_c(p-q)\Delta(q)\Gamma(q|k)\Delta(q-k). \qquad (151)$$

To renormalize these equations one should to add three conter-terms: δm^2 (renormalization of the phion mass), z (the phion-field renormalization) and z_g (the coupling renormalization). In this super-renormalizable model does the only counterterm δm^2 be infinite.

We shall use the normalization at zero momentum

$$\Delta^{-1}(0) = m^2, \quad \left.\frac{d\Delta^{-1}}{dp^2}\right|_{p^2=0} = 1. \qquad (152)$$

Here and below Δ and m are the renormalized quantities.

This zero-momentum normalization condition plays a very important role in the construction of the analytic solutions obtained below. When normalizing at another point, the equations become more complicated, and it is hardly possible to solve them without using numerical methods.

The renormalized equation for the phion propagator becomes

$$\Delta^{-1}(p^2) = m^2 + p^2 + \Sigma_r(p^2), \qquad (153)$$

where $\Sigma_r(p^2) = \Sigma(p^2) - \Sigma(0) - p^2\Sigma'(0)$ and

$$\Sigma(p^2) = -\bar{g}^2 \int \frac{d^4q}{(2\pi)^4} \frac{\Delta(q)}{(p-q)^2}. \qquad (154)$$

Here $\bar{g} \equiv gz_g$, where g is the renormalized coupling.

The vertex also is normalized at zero momenta:

$$\Gamma(0|0) = \frac{g}{\sqrt{N}}, \tag{155}$$

and the renormalized equation for the vertex becomes

$$\Gamma(p|k) = \frac{g}{\sqrt{N}} + \bar{g}^2 \int \frac{d^4q}{(2\pi)^4} \left[\frac{1}{(p-q)^2} \Delta(q)\Gamma(q|k)\Delta(q-k) - \frac{1}{q^2} \Delta(q)\Gamma(q|0)\Delta(q) \right]. \tag{156}$$

After the angle integration with the formula (90) we obtain for the phion propagator the integral equation:

$$\Delta^{-1}(p^2) = m^2 + (1-\kappa)p^2 + 2\kappa m^2 \int_0^{p^2} \Delta(q^2)\left(1 - \frac{q^2}{p^2}\right) dq^2, \tag{157}$$

where

$$\kappa \equiv \frac{\bar{g}^2}{32\pi^2 m^2} \tag{158}$$

is dimensionless coupling.

This equation is reduced to the non-linear differential equation

$$\frac{d^2}{(dp^2)^2}\left(p^2 \Delta^{-1}(p^2)\right) = 2(1-\kappa) + 2\kappa m^2 \Delta(p^2). \tag{159}$$

We shall look for the positive solutions ($\Delta^{-1}(p) > 0$) of the equation for the propagator in the euclidean region of momenta. Negative solutions necessarily contain Landau singularities and are therefore physically unacceptable.

In dimensionless variable $t = p^2/m^2$ and for dimensionless function

$$y(t) = \frac{t}{m^2}\Delta^{-1}$$

the equation (159) becomes

$$y\ddot{y} - 2(1-\kappa)y = 2\kappa t \tag{160}$$

with initial conditions

$$y(0) = 0, \quad \dot{y}(0) = 1. \tag{161}$$

Depending on the value of κ, three different types of positive solutions are possible.

In the weak-coupling region $\kappa < 1$ the asymptotic solution at large p^2 is

$$\Delta^{-1}(p) = (1 - \kappa)p^2 + o(p^2).$$

This asymptotic solution is positive at $\kappa < 1$ and corresponds to the asymptotically-free behavior of propagator.

The approximate analytical solution of equation (160) can be found with the linearization procedure:

$$y = y_0 + y_1, \quad y\ddot{y} \approx y_0\ddot{y}_0 + y_0\ddot{y}_1 + \ddot{y}_0 y_1, \tag{162}$$

and the following conditions on the functions y_0 and y_1 are imposed:
a) y_0 has the right asymptotic behavior and $y_1 = o(y_0)$ at $t \to \infty$;
b) the initial conditions (161) at $t = 0$ are fulfilled.

In the case of the weak coupling one can choose

$$y_0 = (1 - \kappa)t^2 + t, \tag{163}$$

and the equation for y_1 will be following:

$$(1 + (1 - \kappa)t)\ddot{y}_1 = 2\lambda.$$

This equation can be easily integrated, and the solution of the linearized problem in the weak coupling region is

$$\Delta^{-1}(p^2) = (1-\kappa)p^2 + \frac{1-3\kappa}{1-\kappa}m^2 + \frac{2\lambda}{(1-\kappa)^2}\frac{m^2}{p^2}\left(m^2 + (1-\kappa)p^2\right)\log\frac{m^2 + (1-\kappa)p^2}{m^2}. \tag{164}$$

This solution has the right self-consistent asymptotic behavior at $p^2 \to \infty$ and fulfills the initial conditions (161).

At critical coupling $\kappa = 1$ the equation (160) is the singular Emden-Fowler equation

$$\ddot{y} = 2ty^{-1}. \tag{165}$$

Equation (165) has exact solution

$$y_{cr} = \sqrt{\frac{8}{3}}\, t^{3/2}, \tag{166}$$

and the corresponding propagator is

$$\Delta_{cr} = \frac{1}{m}\sqrt{\frac{3}{8p^2}}. \qquad (167)$$

Since Δ_{cr} is the exact solution of equation (157) at $\kappa = 1$ without inhomogeneous term in the r.h.s., then it is asymptotics of the solution of equation (157) at large momenta.

According to this, the asymptotic behavior of propagators in the critical point $\kappa = 1$ has the form

$$\Delta(p) = \sqrt{\frac{3}{8m^2p^2}}\left(1 + O(1/p^2)\right)$$

at large p^2 and drastically differs from the asymptotically-free behavior in the weak-coupling region.

To solve the linearized problem at $\kappa = 1$ one should takes as y_0 the function

$$y_0 = \sqrt{\frac{8}{3}}\,t^{3/2} + t. \qquad (168)$$

With such y_0 the equation for y_1 is

$$t(4t + \sqrt{6t})\ddot{y}_1 + 3y_1 = -3t.$$

The change of variable $x = -\sqrt{8t/3}$ transforms this equation into the hypergeometrical ones, and the solution of the linearized problem in this case is the real-valued positive function

$$y = \sqrt{\frac{8}{3}}\,t^{3/2} + tF\left(1 + i\sqrt{2}, 1 - i\sqrt{2}; 3; -\sqrt{\frac{8t}{3}}\right). \qquad (169)$$

Here F is the Gauss hypergeometrical function [54].

At $\kappa > 1$ equation (160) has the positive exact solution

$$y_s = \frac{\kappa}{\kappa - 1}\,t, \qquad (170)$$

and, correspondingly, equation (159) has solution

$$\Delta_s = \frac{\kappa - 1}{\kappa}\,\frac{1}{m^2}. \qquad (171)$$

Easy to prove that Δ_s is the asymptotic solution of integral equation (157) at $p^2 \to \infty$.

The linearization of equation (160) with $y_0 = \kappa t/(\kappa - 1)$ leads to the equation for y_1:

$$t\ddot{y}_1 + \xi^2 y_1 = 0. \tag{172}$$

Here

$$\xi = (\kappa - 1)\sqrt{\frac{2}{\kappa}}. \tag{173}$$

The solution of equation (172), which fulfills the initial conditions (161), is

$$y_1 = -\frac{\sqrt{t}}{(\kappa - 1)\xi} J_1(2\xi\sqrt{t}),$$

where J_1 is the Bessel function.

Correspondingly, for the inverse propagator we obtain

$$\Delta^{-1}(p^2) = \frac{m^2}{\kappa - 1}\left[\kappa - \frac{1}{\xi}\sqrt{\frac{m^2}{p^2}} J_1\left(2\xi\sqrt{\frac{p^2}{m^2}}\right)\right]. \tag{174}$$

This function is positive at all $p^2 \geq 0$ and gives the true ultraviolet asymptotics and zero-momentum behavior.

Strong-coupling propagator (174) has very interesting shell structure in the four-dimensional Euclidean x–space. Using the Mellin-Barnes representation for J_1 (see, e.g., [55]) one can calculate the Fourier transform of (174):

$$\Delta^{-1}(x) = \int \frac{d^4p}{(2\pi)^4} e^{-ipx} \Delta^{-1}(p) = \frac{\kappa m^2}{\kappa - 1}\left(\delta^4(x) - \frac{m^2}{8\pi^2(\kappa - 1)^2}\delta(x^2 - x_0^2)\right) \tag{175}$$

where the quantity

$$x_0 = (\kappa - 1)\sqrt{\frac{8}{\kappa m^2}}$$

can be considered as the "euclidean radius" of the phion. This radius increases as the coupling increases, which is very natural for the strong–coupling regime.

In the context of the linearized equation considered here it is not clear whether this shell structure an immanent property of the model or is an artifact of linearization. An argument in favor of the first statement is the study of the vertex function.

Consider the equation for the vertex function (156) with zero transfer momentum $k = 0$. After the angle integration with the formula (90) the renormalized equation for the vertex at $k = 0$ becomes

$$\Gamma(p^2) \equiv \Gamma(p|0) = \frac{g}{\sqrt{N}} + 2\kappa m^2 \left\{ \frac{1}{p^2} \int_0^{p^2} q^2 dq^2 \Delta^2(q^2) \Gamma(q^2) - \int_0^{p^2} dq^2 \Delta^2(q^2) \Gamma(q^2) \right\}. \tag{176}$$

This equation can be reduced to differential equation

$$\frac{d^2}{d(p^2)^2} (p^2 \Gamma(p^2)) = -2\kappa m^2 \Delta^2(p^2) \Gamma(p^2) \tag{177}$$

with the initial conditions which follow from the integral equation (176).

In the strong coupling region ($\kappa > 1$) the approximation of the propagator by its asymptotics (171) leads to equation for Γ, which can be reduced to equation (172), and the solution has the form

$$\Gamma(p^2) = \frac{gm}{\xi \sqrt{Np^2}} J_1(\frac{2\xi}{m} \sqrt{p^2}). \tag{178}$$

In x-space we obtain from (178):

$$\Gamma(x) = \int \frac{d^4p}{(2\pi)^4} e^{-ipx} \Gamma(p^2) = \frac{g}{\sqrt{N} x_0^2} \delta(x^2 - x_0^2), \tag{179}$$

i.e., the same shell structure as for the propagator. Note that no linearization was performed in above calculations of Γ. Therefore, it is reasonable to conclude that this shell structure is not an artifact of linearization, but is related to the ultraviolet behavior of the model.

A sharp change of asymptotic behavior in the vicinity of the critical value, which is the main result of this Section, is characteristic for a phase transition. This phenomenon is similar to the re-arrangement of physical vacuum in the strong external field (see [56] – [58] and references therein). As is known the Dirac equation with a potential corresponding to a point charge Ze (Ze is the nuclear charge) is not correct for $Z > 137$: here the "fall on the center" known from quantum mechanics occurs (see, for example, [59], [60]). This fall on the center is related with the term $1/r^2$ in a potential of the relativistic Coulomb problem and is the main reason for this re-arrangement of the vacuum. The potential U, which corresponds to the propagator, is defined by comparing the

Born approximation of the non-relativistic quantum theory with the lower approximation of the relativistic theory (see, e.g., [61], [62]):

$$U(r) \sim -\frac{g^2}{m^2}\phi_{cl}(r), \quad (180)$$

where $r = |\mathbf{x}|$, and ϕ_{cl} is the response of the classical field

$$\phi_{cl}(x) = \int dx_1 \Delta_M(x - x_1) j(x_1) \quad (181)$$

on static source

$$j(x) = \delta^3(\mathbf{x}). \quad (182)$$

In equation (181) Δ_M is the propagator in pseudoeuclidean Minkowski space.

For free propagator $\Delta_c = 1/(m^2 - p^2)$ equations (180) and (181) give the Yukawa potential:

$$U(r) \sim -\frac{g^2}{m^2}\frac{e^{-mr}}{r}.$$

At critical value $\kappa = 1$ for propagator (167) we obtain

$$U(r) \sim -\frac{1}{m}\frac{1}{r^2}. \quad (183)$$

It is a potential of "fall on the center". As we see, the asymptotic behavior of the propagator at the critical point $\kappa = 1$ corresponds to the behavior of the potential at small distances which leads to a phase transition in the supercritical field. Despite all the obvious limitation of this analogy, it undoubtedly indicates the related nature of these phase transitions.

CONCLUSION

Our results demonstrate that two-particle approximation of the system of Schwinger-Dyson equations essentially differs for strictly renormalized models in the asymptotic deep-Euclidean region of momenta in comparison with the standard mean-field approximation. In the theory of a scalar field with quadric self-action in the framework of the two-particle approximation, there is a critical coupling constant λ_{cr}, which divides the whole set of coupling values in two region: the weak-coupling region of inconsistency and the strong-coupling region of non-trivial self-consistent behavior. In the Yukawa model instead of physically unsatisfactory behavior with the Landau pole in the Euclidean region, which occurs for the mean-field approximation, or for the leading term of $1/N$-expansion, the boson propagator in the two-particle approximation has self-consistent asymptotic behavior, which is similar to the asymptotically free behavior. Certainly, inclusion of the scalar self-action into consideration can notably vary the results in the asymptotic region. From this point of view, the obtained results should be considered as the first step of investigation of the full model, which will include Yukawa interaction and the self-action of the scalar field.

In the super-renormalizable scalar Yukawa model the equations of the leading order of $1/N$–expansion have self-consistent positive solutions in the Euclidean region not only in the weak-coupling region, (where a dominance of the perturbation theory in this model is obvious), but also in the strong-coupling region. The phion propagator in the strong-coupling region asymptotically approaches to a constant. It is not something unexpected, if we remember the well-known conception of the static ultra-local approximation, or "static ultralocal model" (see [63] and references therein). In this approximation, all the correlation functions are combinations of δ-functions in the coordinate space that are constants in momentum space. Of course, this approximation is physically trivial. Nevertheless, it can be considered as a starting point for an expansion in inverse powers of the coupling constant, i.e., as a leading approximation of the strong-coupling expansion. In contrast to the ultra-local approximation, our solution has the standard pole behavior for the small momenta. The phenomenon of a sharp change in asymptotic behavior in the considered model can be a quantum-field analogue of re-arrangement of the physical vacuum in a strong external field. In this connection, of interest is the further study of this critical phenomenon, and the search for analogs in other models.

REFERENCES

[1] Kazakov, D.I. (2019). Prospects of elementary particle physics. *Phys. Usp.*, 62: 364-377.

[2] Kazakov, D.I. (2018). Beyond the Standard Model'17. *CERN Yellow Rep. School Proc.*, 3: 83-131.

[3] Mondragon, M. (2018). Beyond the Standard Model. *CERN Yellow Rep. School Proc.*, 4: 101-124.

[4] Landau, L.D., Abrikosov, A.A. and Khalatnikov, I.M. (1954). An asymptotic expression for the photon Green function in quantum electrodynamics. *Dokl. Akad. Nauk Ser. Fiz.*, 95: 1177-1120.

[5] Landau, L. D. and Pomeranchuk, I. Ya. (1955). On point interactions in quantum electrodynamics. *Dokl. Akad. Nauk Ser. Fiz.* 102: 489

[6] Landau, L.D. (1965). *Collected papers of L. D. Landau* (Pergamon Press), Articles 84, 100.

[7] Callaway, D. (1988). Triviality Pursuit: Can Elementary Scalar Particles Exist? *Phys. Reports*, 167: 241-420.

[8] Weisz, P. and Wolff, U. (2011). Triviality of ϕ_4^4 theory: small volume expansion and new data. *Nucl. Phys.*, B846: 316-337.

[9] Jafarov, R.G. and Mutallimov, M.M. (2016). Landau ghost pole problem in quantum field theory: From 50th of last century to the present day. *AIP Conf. Proc.*, 1722: 020003.

[10] Aizenman, M. (1981). Proof of the triviality of ϕ_d^4 field theory and some mean field features of Ising model for $d > 4$. *Phys. Rev. Lett.*, 47: 1-4.

[11] Fröhlich, J., (1982). On the triviality of $\lambda\phi_d^4$ theories and the approach to the critical point in $d \geq 4$ dimensions. *Nucl. Phys.*, B 200: 281-296.

[12] Fernandez, R., Fröhlich, J. and Sokal, A.D. (1992). *Random Walks, Critical Phenomena and Triviality in Quantum Field Theory* (Springer, Berlin).

[13] Weinberg, S. (1995). *The Quantum Theory of Fields Vol. II, Ch. 18* (Cambridge Univ. Press).

[14] Eichhorn, A. (2019). An asymptotically safe guide to quantum gravity and matter. *Front. Astron. Space Sci.*, 5: 47.

[15] Eichhorn, A. and Held, A. (2019). Towards implications of asymptotically safe gravity for particle physics. *Conference*: C19-04-22; *e-Print*: arXiv:1907.05330 [hep-th].

[16] Weinberg, S. (1979). Ultraviolet divergences in quantum theories of gravitation. *General Relativity: Einstein Centenary Survey, eds. S.W. Hawking and W. Israel* (Cambridge University Press, Cambridge). 790-831.

[17] Weinberg, S. (2009). Living with Infinities. *e-Print*: arXiv:0903.0568 [hep-th].

[18] Niedermaier, M. (2007). The asymptotic safety scenario in quantum gravity: An Introduction. *Class. Quant. Grav.*, 24: 171-230.

[19] Pelaggi, G.M., Plascencia, A.D., Salvio, A., Sannino, F., Smirnov, J. and Strumia, A. (2018). Asymptotically Safe Standard Model Extensions? *Phys. Rev.*, D97: 095013.

[20] Barducci, D., Fabbrichesi, M., Nieto, C.M., Percacci, R. and Vedran Skrinjar, V. (2018). In search of a UV completion of the standard model? 378,000 models that don't work. *JHEP*, 1811: 057.

[21] Redmond, P.J. (1958). Elimination of Ghosts in Propagators. *Phys. Rev.*, 112: 1404;

[22] Bogoliubov, N.N., Logunov, A.A. and Shirkov, D.V. (1959). The dispersion relations technique and perturbation theory. *Zh. Eksp. Teor. Fiz.*, 37: 805-815.

[23] Suslov, I.M. (2001). Summing divergent perturbative series in a strong coupling limit. The Gell-Mann-Low function of the ϕ^4 theory. *J. Exp. Theor. Phys.*, 93: 1-23.

[24] Suslov, I.M. (2001). Gell-Mann-Low function in QED. *JETP Lett.*, 74: 191-195.

[25] Suslov, I.M. (2002). Comments on the article by D.I. Kazakov and V.S. Popov *J. Exp. Theor. Phys.*, 95: 601-604.

[26] Suslov, I.M. (2015). A thorny path of field theory: from triviality to interaction and confinement. *e-Print*: arXiv:1506.06128 [hep-ph].

[27] Kazakov, D.I. and Popov, V.S. (2002). On the summation of divergent perturbation series in quantum mechanics and field theory. *J. Exp. Theor. Phys.*, 95: 581-600.

[28] Kazakov, D.I. and Popov, V.S. (2003). Asymptotic behavior of the Gell-Mann-Low function in quantum field theory. *JETP Lett.*, 77: 453-457.

[29] Halpern, K. and Huang, K. (1996). Nontrivial directions for scalar fields. *Phys. Rev.*, D53: 3252-3259.

[30] Djukanovic, D., Gegelia, J. and Meissner, U.-G. (2018). Triviality of quantum electrodynamics revisited. *Commun. Theor. Phys.*, 69: 263.

[31] Rochev, V.E. (2011). Asymptotic behavior in the scalar field theory. *J. Phys. A: Math. Theor.*, 44: 305403.

[32] Rochev, V.E. (2012). Asymptotic behavior in a model with Yukawa interaction from Schwinger-Dyson equations. *J. Phys. A: Math. Theor.*, 45: 205401.

[33] Rochev, V.E. (2013). Asymptotic behavior and critical coupling in the scalar Yukawa model from Schwinger-Dyson equations. *J. Phys. A: Math. Theor.*, 46: 185401.

[34] Rochev, V.E. (2015). System of Schwinger-Dyson equations and asymptotic behavior in the Euclidean region. *Phys. Atom. Nucl.*, 78: 443-446.

[35] Rochev, V.E. (2018). Asymptotic behavior and critical coupling in scalar Yukawa model. *Int. J. Mod. Phys. Conf. Ser.*, 47: 1860095.

[36] Rochev, V.E. (2018). On the phase structure of vector-matrix scalar model in four dimensions. *Eur. Phys. J.*, C78: 927.

[37] Guasch, J., Penaranda, S. and Sanchez-Florit R. (2009). Effective description of squark interactions. *JHEP*, 0904: 016.

[38] Abreu, L.M., Malbouisson, A.P.C., Malbouisson, J.M.C., Nery, E.S. and Rodrigues Da Silva, R. (2014). Thermodynamic behavior of the generalized scalar Yukawa model in a magnetic background. *Nucl. Phys.*, B881: 327-342.

[39] Baym, G. (1960). Inconsistency of Cubic Boson-Boson Interactions. *Phys. Rev.*, 117: 886-888.

[40] Cornwall, J.M. and Morris, D.A. (1995). Toy models of nonperturbative asymptotic freedom in ϕ^3 in six-dimensions. *Phys. Rev.*, D52: 6074-6086.

[41] Savkli, C., Gross, F. and Tjon, J. (2005). Nonperturbative dynamics of scalar field theories through the Feynman-Schwinger representation. *Phys. Atom. Nucl.*, 68: 842-860.

[42] Dahmen, H.D. and Jona-Lasinio, G. (1967). Variational formulation of quantum field theory. *Nuovo Cim.*, A52: 807-836.

[43] Kazanskii, A.K. and Vasilev, A.N. (1972). Legendre transformations for generating functionals in quantum field theory. *Teor. Mat. Fiz.*, 12: 352-369.

[44] Cornwall, J.M., Jackiw, R. and Tomboulis, E. (1974). Effective Action for Composite Operators. *Phys. Rev.*, D10: 2428-2440.

[45] Rochev, V.E. (1997). A Nonperturbative method of calculation of Green functions. *J. Phys. A: Math. Gen.*, 30: 3671-3680.

[46] Rochev, V.E. (2000). On nonperturbative calculations in quantum electrodynamics. *J. Phys. A: Math. Gen.*, 33: 7379-7406.

[47] Nieuwenhuis, T. and Tjon, J.A. (1996). Nonperturbative study of generalized ladder graphs in a $\phi^2\chi$ theory. *Phys. Rev. Lett.*, 77: 814-817.

[48] Ahlig, S. and Alkofer, R. (1999). (In)consistencies in the relativistic description of excited states in the Bethe-Salpeter equation. *Annals Phys.*, 275: 113-147.

[49] Glimm, J. and Jaffe, A. (1987). *Quantum Physics: A Functional Integral Point of View* (Springer, Berlin).

[50] Morris, T.R. (1996). On the fixed point structure of scalar fields *Phys. Rev. Lett.*, 77: 1658.

[51] Rochev, V.E. (2009). Meson contributions in the Nambu-Jona-Lasinio model. *Theor. Math. Phys.*, 159: 488-498.

[52] 't Hooft, G. (1974). A Two-Dimensional Model for Mesons. *Nucl. Phys.*, B75: 461-470.

[53] Slavnov, A.A. (1982). Quantum Chromodynamics And Matrix Models In Terms Of Singlet Variables. 1/n Expansion. *Theor. Math. Phys.*, 51: 517-522.

[54] Luke, Y.L. (1969). *The Special Functions and their Approximations, Vol. 1* (Academic Press, New York).

[55] Erdelyi, A. (1953) *Higher Transcendental Functions, Vol. 2* (McGraw-Hill).

[56] Popov, V.S. (1970). Positron production in a coulomb field with $z > 137$. *Zh. Eksp. Teor. Fiz.*, 59: 965-984.

[57] Greiner, W., Müller, B. and Rafelski, J. (1985). *Quantum Electrodynamics of Strong Fields* (Springer, Berlin).

[58] Kuleshov, V.M., Mur, V.D., Narozhny, N.B., Fedotov, A.M., Lozovik, Yu.E. and Popov, V.S. (2015). Coulomb problem for a $Z > Z_{cr}$ nucleus. *Phys. Usp.*, 58: 785-791.

[59] Landau, L.D. and Lifshitz, E.M. (1989). *Quantum mechanics* (Pergamon Press, N.Y.).

[60] Oksak, A.I.(1986). Scattering In Conformal Invariant Quantum Mechanics. *Theor. Math. Phys.*, 66: 142-146.

[61] Bethe, H. and Morrison, P. (1956). *Elementary Nuclear Theory* (Wiley, N.Y.).

[62] Gross, F. (1999). *Relativistic Quantum Mechanics and Field Theory* (Wiley, N.Y.).

[63] Rivers, R.J. (1987). *Path Integral Methods in Quantum Field Theory* (Cambridge Univ. Press).

In: Asymptotic Behavior: An Overview
Editor: Steve P. Riley

ISBN: 978-1-53617-222-5
© 2020 Nova Science Publishers, Inc.

Chapter 3

ASYMPTOTIC BEHAVIOR FOR THE HYDROGEN ATOM CONFINED BY DIFFERENT POTENTIALS

Michael-Adán Martínez-Sánchez, Rubicelia Vargas and Jorge Garza[*]
Departamento de Química
Universidad Autónoma Metropolitana-Iztapalapa
Iztapalapa, México City, México

Abstract

The exact wave function of the *free* hydrogen atom has been used to design basis sets employed for the study of many-electron atoms; however, when we take into account the impact of the environment in which is immersed an atom, the behavior of the wave function is different from that showed by the free atom. In this chapter, we discuss the exact analytical solution of the Schrödinger equation corresponding to the hydrogen atom confined by four spherical potentials: I) Infinite potential. II) Parabolic potential. III) Constant potential. IV) Dielectric continuum. All these potentials are applied in the region $r \geq r_0$ where r_0 constitutes a confinement radius. In all cases, the potential $-Z/r$ is defined in the region $r < r_0$. As a general conclusion we found that the wave function decays faster in the following ordering: Pot. I > Pot. II > Pot. III > Pot. IV. The analytical expression for the exact wave function shows how the

---
[*]Corresponding Author's E-mail: jgo@xanum.uam.mx.

asymptotic behavior is for each potential and consequently this information suggests how to build a basis set to study the electronic structure of many-electron atoms under the same spatial restrictions. Thus, we present results when Gaussian functions are used as a basis set in the Ritz method and we contrast its results with those obtained by the exact solution for the confined hydrogen atom.

Keywords: correct asymptotic behavior, confined atoms, exact solution, ground and excited states, penetrable and impenetrable walls

1. Introduction

The hydrogen atom is a cornerstone in quantum mechanics since this system is one of the first applications given by Schrödinger to show the importance of his equation. [1] With the analytical solution of this system, quantum chemistry people proposed the Ritz method to study the electronic structure of many-electron atoms and molecules through Slater type orbitals (STO) defined as

$$R_{\text{STO}}(r) = N r^{n-1} e^{-\zeta r}, \qquad (1)$$

where N represents a normalization constant, n is a positive integer and ζ is an exponent that minimizes the total energy. John C. Salter based his proposal on "the asymptotic form at large distances for a hydrogen-like wave function" [2]. Without a doubt this proposal was the key for the development of quantum chemistry. Thus, the analytical solution of the hydrogen atom provided one way to build basis sets for atoms and molecules. Currently, many quantum chemistry codes use Gaussian type orbitals (GTO) [3]

$$R_{\text{GTO}}(r) = N r^{n-1} e^{-\alpha r^2}. \qquad (2)$$

The main reason to use GTOs instead STOs is the numerical work involved in the evaluation of two-electron integrals for molecular systems [4, 5, 6], since there are recurrence relations with GTOs that alleviate the computational effort. However, for atomic systems STOs are the best alternative to represent Hartree-Fock [7] or Kohn-Sham [8] orbitals in quantum chemistry.

In addition to the *free* hydrogen atom, there are several reports where this system is confined under different spatial restrictions and the corresponding Schrödinger equation is solved analytically. The study of the confined hydrogen atom, in general, has been focused on this atom centered within a spherical cavity of radius r_0 with a potential, in atomic units (au), defined as

$$v(r) = \begin{cases} -\dfrac{Z}{r} & 0 \leq r < r_0 \\ V_c & r_0 \leq r < \infty \end{cases}, \qquad (3)$$

where Z is the atomic number and V_c represents a confinement potential. Thus, within the cavity ($r < r_0$) the potential corresponds to the nucleus-electron interaction, and for $r \geq r_0$ the potential is designed to simulate different physical environments. The first confinement proposed to simulate high pressures over the hydrogen atom is related to the potential

$$V_c = \infty. \qquad (4)$$

Since the confinement potential is infinite, the wave function and the corresponding electron density are canceled on the surface of the cavity, for that reason this model is named in several ways, as impenetrable walls model or hard walls model. Such a model was introduced by Michels et al. [9], these authors applied this model on the hydrogen atom to simulate extreme pressures, and they found a reasonable comparison between their results and experimental data. To approximate this problem, with many-electron atoms as main target, Ludeña proposed as basis set [10, 11]

$$R_{\text{Imp}}(r) = R_{\text{STO}}(r)\left(1 - \dfrac{r}{r_0}\right). \qquad (5)$$

In this way, the basis set for impenetrable walls is a STO times a cutoff function, which ensures that the wave function is zero at $r = r_0$. This basis set has been used to study several properties of many-electron atoms by using wave function techniques [11, 12, 13, 14, 15]. In the best of our knowledge this basis set has been used only by our group to study the electronic structure through the density functional theory [16, 17].

The impenetrable walls model gives an idea about atoms under extreme pressures, unfortunately, the pressure is overestimated since the total energy grows up rapidly for small confinement radii. For that reason, penetrable potentials have been reported to alleviate the mentioned issue involved with impenetrable walls [18]. Ley-Koo and Rubinstein proposed a penetrable walls model as [19]

$$V_c = U_0, \qquad (6)$$

where U_0 is a constant and in consequence, constitutes a well of finite height. Such a potential was useful to reproduce experimental pressures over the helium atom [18]. The analytical solution for the hydrogen atom under this confinement shows that the asymptotic behavior of the wave function must be

$$R_{\text{fnt}}(r) \sim r^{-\ell-1}e^{-\kappa r}, \qquad (7)$$

with ℓ as the angular-momentum quantum number and κ is a parameter related to the constant potential U_0. From this expression we observe that for the lowest value of the angular moment ($\ell = 0$) we have

$$R_{\text{fnt}}(r) \sim r^{-1}e^{-\kappa r}. \qquad (8)$$

This result shows that the wave function will be canceled faster than the *free* atom for high values of r, which has sense since the corresponding wave function is attenuated in classically forbidden regions.

The analytical solution expressed in equation (7) inspired to Rodriguez-Bautista et al. to suggest the basis set [20]

$$R_{\text{STO-m}}(r) = \begin{cases} R_{\text{STO}}(r) & r < r_0, \\ N_{\text{out}} r^{-\ell-1} e^{-\alpha r} & r \geq r_0. \end{cases} \qquad (9)$$

Such a basis set was used successfully to solve Hartree-Fock and Kohn-Sham equations for many-electron atoms confined by penetrable walls [20, 16], where some properties present a behavior quite different than that observed for atoms confined by impenetrable walls or for *free* atoms. We want to mention an interesting result observed for this confinement; the confinement induces an ionization process in an atom when the confinement radius acquires small values. Right before to this critical confinement radius, r_{crit}, the electron is spread over the classically forbidden region, which has been measured through the Shannon entropy in configuration space [21].

For many years, a dielectric continuum has been used to model solvent effects on atoms and molecules [22, 23]. In this solvent model, atoms or molecules are inside of a cavity within a dielectric continuum, in other words, the dielectric confines atoms or molecules, which has been recognized from early stages of this model [24]. For this case, the confinement is modeled as

$$V_c = -\frac{Z}{\epsilon r}, \qquad (10)$$

where ϵ is the relative permittivity. This confinement imposes penetrable walls in a different way to that proposed by Ley-Koo and Rubinstein through equation (6) and consequently the wave function associated to this problem is different than that presented in equation (7). In particular, from the exact analytical solution of the hydrogen atom immersed in a dielectric continuum, Martinez-Sánchez et al. found that the asymptotic behavior for this confinement is [25]

$$R_{\text{dlc}}(r) \approx N r^\beta e^{-\alpha r}, \tag{11}$$

where α and β are real numbers (β is not necessarily an integer) and therefore this solution does not correspond to a STO, and this is not the same to that found by Ley-Koo and Rubinstein for a constant potential.

The confinements presented at this point can be divided into two groups: 1) Finite potentials and, 2) infinite potentials. The finite potentials impose penetrable walls and the infinite potential mentioned in this chapter impose impenetrable walls. However, the potential

$$V_c = \frac{1}{2}\omega^2 r^2, \tag{12}$$

is infinite when $r \to \infty$, although unlike the infinite potential, this potential is increased gradually in this limit and therefore it has associated penetrable walls. Thus, it is interesting the study of this confinement to compare its results with those obtained by other confinement potentials. It is worth to note that such a comparison has not been yet reported elsewhere.

In this chapter we discuss an approach to solve analytically the Schrödinger equation for the hydrogen atom confined by the four potentials mentioned above and we compare the results generated by all these confinements. Finally, with the analytical solution of the Schrödinger equation for the hydrogen atom confined by four potentials we do build a basis set of GTOs to analyze its performance concerning exact results.

2. EXACT ANALYTICAL SOLUTION OF THE SCHRÖDINGER EQUATION

The exact solution of the Schrödinger equation for the hydrogen atom confined by a constant potential was provided by Ley-Koo and Rubinstein [19]. Martínez-Sánchez et al. followed such a procedure to find the exact solution of the confinement imposed by a dielectric continuum. This procedure can be summarized in the following recipe: **A)** Write the wave function as a power series within the region $r < r_0$. **B)** Write the wave function in terms of confluent hypergeometric functions in the region $r \geq r_0$. **C)** Impose boundary conditions to find the r_0 which corresponds to a fixed energy. Such a technique has been employed to solve the problems associated to equations (6) and (10) and it will be applied to solve the problem of the parabolic potential.

The four confinements considered in this chapter exhibit spherical symmetry and consequently the solution can be written as a product of a radial part and an angular contribution. The spherical harmonics functions are solutions of the angular part and the radial equation, in au is given by

$$\left[-\frac{1}{2}\frac{\partial^2}{\partial r^2} - \frac{1}{r}\frac{\partial}{\partial r} + \frac{\ell(\ell+1)}{2r^2} + v(r)\right] R(r) = \xi R(r), \tag{13}$$

where ξ is the total energy of the system and $v(r)$ is the potential defined in equation (3).

2.1. Solution within the Cavity ($r < r_0$)

Inside the cavity, the differential equation to solve is

$$\left[-\frac{1}{2}\frac{\partial^2}{\partial r^2} - \frac{1}{r}\frac{\partial}{\partial r} + \frac{\ell(\ell+1)}{2r^2} - \frac{Z}{r}\right] R_{in}(r) = \xi R_{in}(r). \tag{14}$$

The solution of this equation is written as [26]

$$R_{in}(r) = A r^\ell \Phi(r), \tag{15}$$

where A is a normalization constant and $\Phi(r)$ must satisfy

$$\left[r\frac{\partial^2}{\partial r^2} + 2(\ell+1)\frac{\partial}{\partial r} + 2Z + 2\xi r\right] \Phi(r) = 0. \tag{16}$$

Proposing $\Phi(r)$ as a series expansion,

$$\Phi(r) = \sum_{i=0}^{\infty} c_i r^i, \tag{17}$$

the coefficients must satisfy a three terms recurrence relation

$$c_{i+2} = -\frac{2(Zc_{i+1} + \xi c_i)}{(i+2)(i+2\ell+3)}, \tag{18}$$

with

$$c_1 = -\frac{Z}{\ell+1}c_0. \tag{19}$$

By convenience, $c_0 = 1$ without loss of generality since the normalization is imposed over A. At this point, the solution is the same within this region for the four potentials considered in this chapter.

2.2. Solution outside the Cavity ($r \geq r_0$)

In this interval of r each confinement potential yields a different form for the differential equation to find the corresponding analytical solution. However, with an appropriate transformation we can obtain the Kummer's equation [27],

$$\left[x\frac{\partial^2}{\partial x^2} + (b-x)\frac{\partial}{\partial x} - a \right] f(x) = 0. \tag{20}$$

This differential equation has two linearly independent solutions reported in the literature as $M(a, b, x)$ (confluent hypergeometric function of the first kind) and $U(a, b, x)$ (confluent hypergeometric function of the second kind). For an infinite potential this equation is not valid since for this case the wave function is zero in this region. Let us obtain the corresponding transformation for each potential to obtain equation (20).

Constant Potential ($V_c = U_0$)

For this case, the confinement potential is a constant and the differential equation to solve is

$$\left[-\frac{1}{2}\frac{\partial^2}{\partial r^2} - \frac{1}{r}\frac{\partial}{\partial r} + \frac{\ell(\ell+1)}{2r^2} + U_0 \right] R_{\text{fnt}}(r) = \xi R_{\text{fnt}}(r). \tag{21}$$

Here, the solution found by E. Ley Koo and S. Rubinstein is written as [19]

$$R_{\text{fnt}}(r) = B_{\text{fnt}} r^{-\ell-1} e^{-\kappa r} f(r), \tag{22}$$

where B_{fnt} is a normalization constant in this region and the parameter κ is defined as

$$\kappa = \sqrt{2(U_0 - \xi)}, \tag{23}$$

such that $f(r)$ must satisfy

$$\left[r\frac{\partial^2}{\partial r^2} - 2(\ell + \kappa r)\frac{\partial}{\partial r} + 2\ell\kappa \right] f(r) = 0. \tag{24}$$

To obtain the Kummer's equation, it is convenient to use the following scaling

$$x = 2\kappa r. \tag{25}$$

In this way equation (24) is transformed in (20) with

$$a = -\ell, \tag{26}$$

and

$$b = -2\ell. \tag{27}$$

For this potential, it is useful to use the solution $f(x) = M(a, b, x)$, which in this case becomes as a polynomial of degree ℓ; i.e. $f(x) = P_\ell(x)$. For example,

$$P_\ell(x) = \begin{cases} 1, & \ell = 0, \text{ s states} \\ 1 + \dfrac{1}{2}x, & \ell = 1, \text{ p states} \\ 1 + \dfrac{1}{2}x + \dfrac{1}{12}x^2, & \ell = 2, \text{ d states} \\ 1 + \dfrac{1}{2}x + \dfrac{1}{10}x^2 + \dfrac{1}{120}x^3, & \ell = 3, \text{ f states} \end{cases} \tag{28}$$

As we mentioned previously, the asymptotic behavior of the radial wave function for this potential is

$$R_{\text{fnt}}(r) \sim r^{-\ell-1} e^{-\sqrt{2(U_0-\xi)}\,r}. \tag{29}$$

In this case, bound states lie in the energy spectrum

$$E_n^{\text{free}} < \xi < U_0, \tag{30}$$

where E_n^{free} is the free hydrogen atom energy. If $\xi > U_0$, the electron is no longer bound to the nucleus and its behavior is like a free particle.

Dielectric Continuum ($V_c = -\frac{Z}{\epsilon r}$)

The differential equation associated to this potential has the form

$$\left[-\frac{1}{2}\frac{\partial^2}{\partial r^2} - \frac{1}{r}\frac{\partial}{\partial r} + \frac{\ell(\ell+1)}{2r^2} - \frac{Z}{\epsilon r}\right] R_{\text{dlc}}(r) = \xi R_{\text{dlc}}(r). \tag{31}$$

In principle, equations (14) and (31) are practically the same, however, in this region $R_{\text{dlc}}(r) \longrightarrow 0$ as $r \longrightarrow \infty$, for that reason Martínez-Sánchez et al. suggested the following solution [25]

$$R_{\text{dlc}}(r) = B_{\text{dlc}} r^\ell e^{-\sqrt{-2\xi}\,r} f(r). \tag{32}$$

Here $f(r)$ must satisfy

$$\left[r\frac{\partial^2}{\partial r^2} + 2\left(\ell + 1 - \sqrt{-2\xi}\,r\right)\frac{\partial}{\partial r} + 2\left(\frac{Z}{\epsilon} - (\ell+1)\sqrt{-2\xi}\right)\right] f(r) = 0. \tag{33}$$

To solve this equation, the next scaling is convenient

$$x = 2\sqrt{-2\xi}\,r, \tag{34}$$

in consequence, equation (33) is transformed in (20) with

$$a = \ell + 1 - \frac{Z}{\epsilon\sqrt{-2\xi}}, \tag{35}$$

and

$$b = 2(\ell + 1). \tag{36}$$

Contrary to the constant potential, in this case $f(x) = U(a,b,x)$ is the solution to be used. For this potential, the radial wave function exhibits the asymptotic behavior

$$R_{\text{dlc}}(r) \sim r^{-\left(1 - \frac{Z}{\epsilon\sqrt{-2\xi}}\right)} e^{-\sqrt{-2\xi}\,r}. \tag{37}$$

As in the previous case, the asymptotic behavior of the wave function is governed by an exponential; however, for this potential the exponent of the pre-exponential factor is not necessarily an integer and also depends on the energy ξ and the relative permittivity ϵ.

For bound states, the energy spectrum in which we are interested is

$$E_n^{\text{free}} < \xi < 0. \tag{38}$$

Parabolic Potential ($V_c = \frac{1}{2}\omega^2 r^2$)

In the best of our knowledge this is the first time that the hydrogen atom is confined by a parabolic potential coupled with a Coulomb interaction, which could be useful to simulate a plasma [28]. For this case, the differential equation to solve is

$$\left[-\frac{1}{2}\frac{\partial^2}{\partial r^2} - \frac{1}{r}\frac{\partial}{\partial r} + \frac{\ell(\ell+1)}{2r^2} + \frac{1}{2}\omega^2 r^2\right] R_{\text{prb}}(r) = \xi R_{\text{prb}}(r), \quad (39)$$

For the solution of this differential equation, we propose

$$R_{\text{prb}}(r) = B_{\text{prb}} r^\ell e^{-\frac{1}{2}\omega r^2} f(r). \quad (40)$$

This proposal gives a differential equation for $f(r)$

$$\left[r\frac{\partial^2}{\partial r^2} + 2\left(\ell + 1 - \omega r^2\right)\frac{\partial}{\partial r} + [2\xi - \omega(2\ell+3)]r\right] f(r) = 0. \quad (41)$$

In this case we use a non-linear scaling,

$$x = \omega r^2, \quad (42)$$

to transform equation (41) in equation (20) with

$$a = \frac{1}{2}\left(\ell + \frac{3}{2} - \frac{\xi}{\omega}\right), \quad (43)$$

and

$$b = \ell + \frac{3}{2}. \quad (44)$$

For this confinement $f(x) = U(a, b, x)$ is the best choice for the solution, which is responsible of the asymptotic behavior of the wave function

$$R_{\text{prb}}(r) \sim r^{-\frac{1}{2}\left(3 - \frac{2\xi}{\omega}\right)} e^{-\frac{1}{2}\omega r^2}. \quad (45)$$

For all potentials considered in this chapter, this is the potential where the asymptotic behavior of the wave function is commanded by a Gaussian function.

To have an idea about the energy spectrum of bound states, we have to consider the limit $r_0 \longrightarrow 0$. When we reach this limit, there is no more Coulomb contribution and; in consequence, the electron only experiences the action of the parabolic potential; therefore, the energy spectrum of bound states must be

$$E_n^{\text{free}} < \xi < E_{n',\ell}^{\text{osc}} \tag{46}$$

where $E_{n',\ell}^{\text{osc}}$ is the energy of an isotropic harmonic oscillator

$$E_{n',\ell}^{\text{osc}} = \left(4n' + 2\ell + 3\right)\frac{\omega}{2}, \tag{47}$$

and $n' = 0, 1, 2, 3, \ldots$ is a quantum number related to the number of nodes of the radial wave function of the isotropic harmonic oscillator. Therefore $n' \neq n$.

2.3. Imposing Boundary Conditions

For potentials defined in two regions, the solution and its first derivative must be continuous at $r = r_0$,

$$R_{\text{in}}(r)\Big|_{r=r_0} = R_{\text{out}}(r)\Big|_{r=r_0}, \tag{48}$$

and

$$\frac{\partial R_{\text{in}}(r)}{\partial r}\bigg|_{r=r_0} = \frac{\partial R_{\text{out}}(r)}{\partial r}\bigg|_{r=r_0}. \tag{49}$$

The first condition gives a relation between normalization constants, and the second condition (using logarithmic derivatives) provides one way to obtain the confinement radius r_0. Let us apply these conditions on equations (15), (22), (32) and (40).

Finite Potential ($V_c = U_0$)

First condition

$$B_{\text{fnt}} = A\frac{r_0^{2\ell+1}e^{\kappa r_0}\Phi(r_0)}{P_\ell(x_0)}. \tag{50}$$

Second condition

$$r_0\left[\Phi'(r_0)P_\ell(x_0) - \Phi(r_0)P_\ell'(x_0)\right] + (2\ell + 1 + \kappa r_0)\,\Phi(r_0)P_\ell(x_0) = 0. \tag{51}$$

where $x_0 = 2\kappa r_0$ and κ is defined in equation (23).

Continuum Dielectric ($V_c = -\frac{Z}{\epsilon r}$)

First condition

$$B_{\text{dlc}} = A \frac{e^{\sqrt{-2\xi}r_0}\Phi(r_0)}{U(a,b,x_0)}. \tag{52}$$

Second condition

$$\left(\Phi'(r_0) + \sqrt{-2\xi}\Phi(r_0)\right)U(a,b,x_0) + 2\left((\ell+1)\sqrt{-2\xi} - \frac{Z}{\epsilon}\right)\Phi(r_0)U(a+1,b+1,x_0) = 0. \tag{53}$$

This equation was obtained by using the property

$$\frac{\partial}{\partial x}U(a,b,x) = -aU(a+1,b+1,x). \tag{54}$$

Here $x_0 = 2\sqrt{-2\xi}r_0$ and the parameters a and b are defined in equations (35) and (36), respectively.

Parabolic Potential ($V_c = \frac{1}{2}\omega^2 r^2$)

First condition

$$B_{\text{prb}} = A \frac{e^{\frac{1}{2}\omega r_0^2}\Phi(r_0)}{U(a,b,x_0)}. \tag{55}$$

Second condition

$$\left(\Phi'(r_0) + \omega r_0 \Phi(r_0)\right)U(a,b,x_0) + 2a\omega r_0 \Phi(r_0)U(a+1,b+1,x_0) = 0. \tag{56}$$

Here $x_0 = \omega r_0^2$ and the parameters a and b are defined in equations (43) and (44), respectively.

For equations (51), (53) and (56) we have used the notation

$$\Phi'(r_0) = \left.\frac{\partial\Phi(r)}{\partial r}\right|_{r=r_0}. \tag{57}$$

Finally, a complete description of the radial wave function is obtained when the normalization constants satisfy the condition

$$\int_0^{r_0} r^2 R_{\text{in}}(r)^2 dr + \int_{r_0}^{\infty} r^2 R_{\text{out}}(r)^2 dr = 1. \tag{58}$$

Impenetrable Wall

This is the only case where $R_{out}(r) = 0$, therefore, we need to calculate only the zeros of the power series by fixing a finite number of terms (n_T), i.e.

$$\Phi(r) = \sum_{i=0}^{n_T} c_i r^i = 0. \tag{59}$$

3. RESULTS

To obtain the solution of the hydrogen atom confined by the four potentials considered in this chapter we fixed Z, ℓ and U_0 for the constant potential, ϵ for the dielectric continuum and ω for the parabolic potential. With this information there are two ways to carry out numerical calculations:

1. Fix ξ to obtain the confinement radius, r_0.

2. Fix r_0 to obtain the corresponding energy, ξ.

In this chapter we use both options, thus we solved equations (51), (53), (56), and (59) by using Mathematica v11.2 [29] with $n_T = 26$ for the series expansion in equation (15).

3.1. Ground State of the Confined Hydrogen Atom

3.1.1. Total Energy

In this section, we present the total energy and its contributions for the ground state ($n = 1, \ell = 0$) of the hydrogen atom confined by the four potentials considered in this chapter. As we mentioned above, in our approach, the total energy is the input to obtain the corresponding confinement radius, or vice versa, r_0 is the entry to obtain ξ. For both cases, the corresponding results are presented in Tables 1 and 2. In both tables we used two values for each potential: a) $\epsilon = 2.5$ and 80.0, b) $U_0 = 0.0$ and 0.5 au, c) $\omega = 0.5$ and 1.0 au. Results related to hard walls are also included in these tables. The impact of the potential imposed on the confinement radius is evidenced in Table 1. To build this table we fixed the energy and we found the corresponding confinement radius.

From here we do observe that if the barrier height is large then the confinement radius is also large. In other words, in a confinement process the energy responds quickly even for large r_0 when the corresponding confinement potential

Table 1. Confinement radius versus total energy for the hydrogen atom confined by four potentials. Energy (ξ), confinement radius (r_0), constant potential (U_0) and ω in atomic units. In this case ξ is the input to solve the corresponding Kummer's equation and r_0 is the solution

	Confinement radius (r_0)						
	Dielectric continuum (U_0)		Constant potential (ϵ)		Parabolic potential (ω)		Inf. potential
ξ	2.5	80.0	0.0	0.5	0.5	1.0	
-0.490	2.90788	3.13972	3.14529	3.56504	3.88423	4.10496	4.33779
-0.400	1.53269	1.77460	1.78020	2.08718	2.17514	2.46196	2.81127
-0.300	1.10203	1.37122	1.34720	1.65794	1.66654	1.95474	2.37974
-0.200	0.80832	1.13235	1.13899	1.41510	1.38330	1.65991	2.13197
-0.125	0.55933	0.99245	1.00000	1.28408	1.23305	1.49926	2.00000
-0.085	0.23426	0.92241	0.93078	1.22506	1.16603	1.42684	1.94123

exhibits high values. Thus, for a fixed energy $r_0^{\text{infty}} > r_0^{\text{parab}} > r_0^{\text{const}} > r_0^{\text{dielec}}$. A summary of this result is presented in Figure 1 where we observe that impenetrable walls increase rapidly the total energy. Contrary to this result, the confinement imposed by a dielectric continuum gives the smallest changes between these potentials. Thus, hard walls cannot be used to simulate the behavior obtained when a dielectric continuum potential is used to mimic solvent effects. Constant and parabolic potentials give confinement radii between those obtained by the infinite potential and the dielectric continuum. For the infinite potential, the total energy grows up infinitely when the confinement radius is reduced. However, the other three potentials present upper bounds for this property. For the dielectric continuum the upper bound is zero (see equation 38), for a constant potential this upper bound is precisely U_0 (see equation 30) and for the parabolic potential the upper bound is that imposed by the harmonic oscillator (see equation 46) since when the confinement radius is zero the potential becomes the harmonic oscillator. These upper bounds are presented as horizontal lines in Figure 1.

From this figure we appreciate that upper bounds for the constant potential and the dielectric continuum are reached for $r_0 = 0.644281$ au and $r_0 = 0.618786$ au, respectively. It means that for these confinement radii the hydrogen atom is ionized.

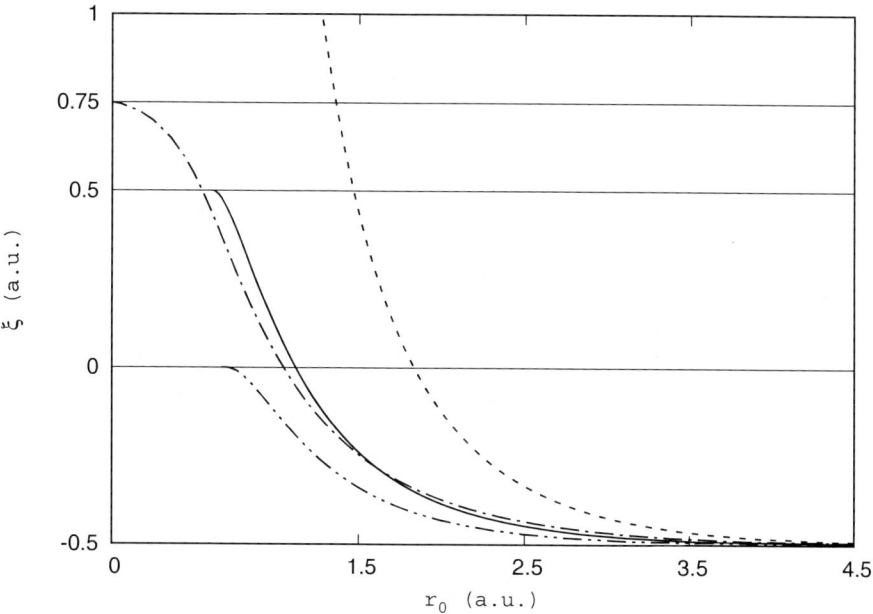

Figure 1. Total energy (ξ) as a function of the radius of the cavity (r_0) for the hydrogen atom confined by impenetrable walls (dashed line), parabolic potential (dot-dashed line $\omega = 0.5$ au), constant potential (solid line $U_0 = 0.5$ au), and dielectric continuum (dot-dot-dashed line $\epsilon = 80$).

In Table 2 the confinement radius, r_0 was the input of the Kummer's equation and the energy represents the solution. This approach is similar to draw a vertical line in the plot of Figure 1. According to the previous paragraph, if we fix r_0 then the infinite potential delivers the highest energy. It is worth noting that there are small values for r_0 where there is no solution for the Kummer's equation since for these confinement radii the hydrogen atom is already ionized.

3.1.2. Components of the Energy

By construction, several properties must be evaluated in two regions since the wave function of the hydrogen atom is defined by $R_{in}(r)$ for $r < r_0$, and $R_{out}(r)$

Table 2. Total energy for the ground state of the hydrogen atom within a sphere of radius r_0 confined by four potentials. Energy (ξ), confinement radius (r_0), constant potential (U_0) and ω in atomic units. In this case r_0 is the input to solve the corresponding Kummer's equation and ξ is the solution

	Total energy (ξ)						
	Dielectric continuum (U_0)		Constant potential (ϵ)		Parabolic potential (ω)		Inf. potential
r_0	2.5	80.0	0.0	0.5	0.5	1.0	
1.0	-0.26720	-0.12927	-0.12500	0.11097	0.03531	0.26516	2.37399
2.0	-0.45407	-0.43186	-0.43122	-0.38511	-0.37348	-0.31202	-0.12500
3.0	-0.49144	-0.48734	-0.48722	-0.47592	-0.46660	-0.45225	-0.42397
4.0	-0.49847	-0.49770	-0.49767	-0.49496	-0.49150	-0.48836	-0.48327
5.0	-0.49973	-0.49959	-0.49958	-0.49899	-0.49800	-0.49735	-0.49641

for $r \geq r_0$. Thus, the kinetic energy (KE) is obtained from two integrals

$$\text{KE} = \int_0^{r_0} dr\, r^2 R_{\text{in}}^*(r) \left(-\frac{1}{2}\frac{\partial^2}{\partial r^2} - \frac{1}{r}\frac{\partial}{\partial r}\right) R_{\text{in}}(r)$$
$$+ \int_{r_0}^{\infty} dr\, r^2 R_{\text{out}}^*(r) \left(-\frac{1}{2}\frac{\partial^2}{\partial r^2} - \frac{1}{r}\frac{\partial}{\partial r}\right) R_{\text{out}}(r). \quad (60)$$

We present this expression to show that KE has two components, one of them uses the wave function in the classically forbidden region $r \geq r_0$, which is interesting since the corresponding integral will be negative or zero for that region.

The kinetic energy as a function of the confinement radius is presented in Figure 2 for each confinement considered in this chapter. This figure shows interesting features of KE when the hydrogen atom is submitted to different confinements. For example, the confinement imposed by impenetrable walls increases KE always when the confinement radius is reduced. In fact, for this confinement the second integral does not appear in equation (60). This energy component exhibits a different behavior for penetrable potentials since in these cases KE reaches a maximum value and acquires smaller values than those delivered by the free atom for small confinement radii. For the parabolic potential, in the limit $r_0 \to 0$ the KE corresponds to that exhibited by the harmonic oscil-

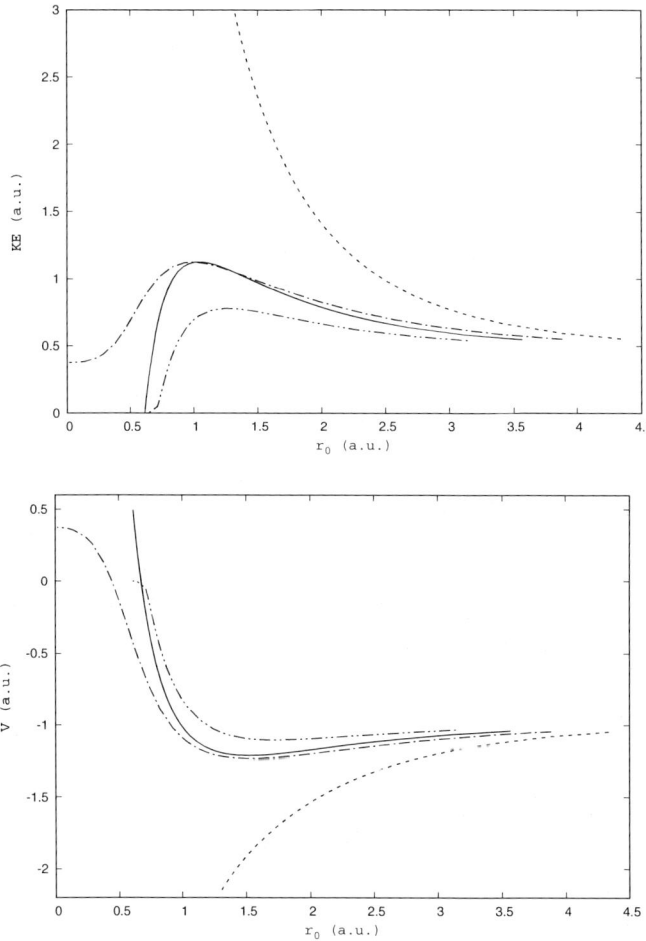

Figure 2. Kinetic energy (KE) and potential energy (V) as functions of r_0 for the hydrogen atom confined by impenetrable walls (dashed line), parabolic potential (dot-dashed line $\omega = 0.5$ a.u.), constant potential (solid line $U_0 = 0.5$ a.u.), and dielectric continuum (dot-dot-dashed line $\epsilon = 80$).

lator. For constant and dielectric continuum potentials, the kinetic energy is zero when the confinement radius reach a critical value, precisely when the electron is detached by the confinement.

The potential energy (V) has two contributions, as the KE discussed previously,

$$V = \underbrace{-Z \int_0^{r_0} dr\, r R_{\text{in}}^*(r) R_{\text{in}}(r)}_{V_{\text{in}}(r_0)} + \underbrace{\int_{r_0}^{\infty} dr\, r^2 R_{\text{out}}^*(r) V_c(r) R_{\text{out}}(r)}_{V_{\text{out}}(r_0)}. \quad (61)$$

The first integral is always negative and the second one depends on the confinement potential V_c. The upper limit of this property is determined by the highest value of the corresponding potential; 0 au for the dielectric continuum, 0.5 au for a constant potential with height barrier of 0.5 au and 0.375 au for the parabolic potential. Hard walls decrease always the potential energy and consequently $V \to -\infty$ when $r_0 \to 0$, as we can observe from Figure 2 where the potential energy is presented for each confinement considered in this chapter.

As a summary of these results, we can say that KE and V delivered by hard walls represent upper and lower limits, respectively, of the hydrogen atom confined by several confinements.

3.1.3. Charge in Classically Forbidden Regions and Electron Density at the Origin

At this point, the discussion has been centered on total energy and its components. In these paragraphs we analyze the electron density of the confined hydrogen atom. The penetration of the electron density inside classically forbidden regions can be obtained through the evaluation of the charge within this region by using the integral

$$Q(r_0) = \int_{r_0}^{\infty} dr\, r^2 R_{\text{out}}(r)^2. \quad (62)$$

This quantity was evaluated for the three penetrable potentials included in this chapter. For an impenetrable potential this quantity has non sense since by definition this potential does not allow penetration of the wave function.

Results for $Q(r_0)$ are presented in Figure 3, where this quantity has been multiplied by 100 to give a percent of the wave function penetration. This plot gives a good idea about the softness exhibited by each potential. Without a doubt the dielectric continuum impose very soft walls since an important percent of the wave function is inside of the confinement potential for regular confinement radii. The charge within classically forbidden regions is quite similar between

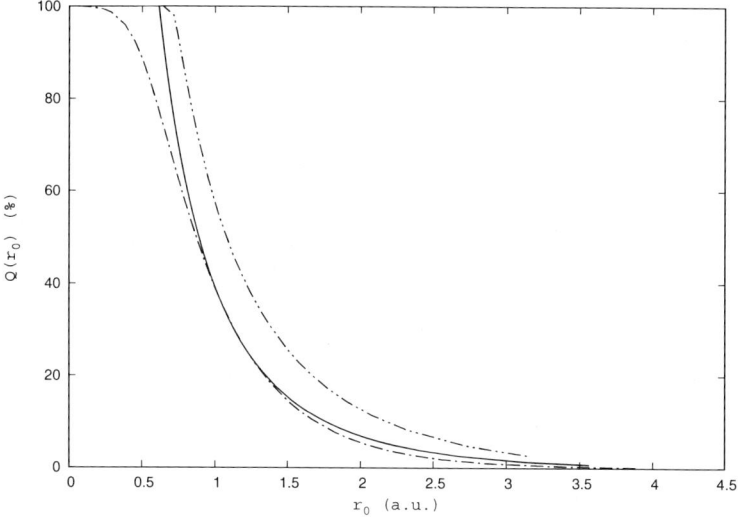

Figure 3. Probability, in percent, to find an electron inside a confinement potential imposed over the hydrogen atom. Constant potential (solid line $U_0 = 0.5$ au), continuum dielectric (dot-dot-dashed line $\epsilon = 80$) and parabolic potential (dot-dashed line $\omega = 0.5$ au).

a constant potential and a parabolic potential for moderate confinements. However, for small confinement radii there is one confinement radius where Q is increased rapidly by the constant potential and reaches the 100 % of penetration. The parabolic potential is harder than the dielectric continuum and the constant potential because we need small confinement radii to reach 95 % or more of penetration.

The electron density evaluated at the origin, $\rho(0)$ is depicted in Figure 4 for each potential considered in this chapter. From this figure, it is evident that a confinement imposed by impenetrable walls always increases this property when the confinement radius decreases, this is a consequence of the localization or concentration of the electron density since it cannot escape from the cavity. However, for other potentials, where the electron density can explore regions outside of the cavity, $\rho(0)$ reaches a maximum at r_0^{max}. For confinement radii less than r_0^{max} this property goes down rapidly for constant and dielectric con-

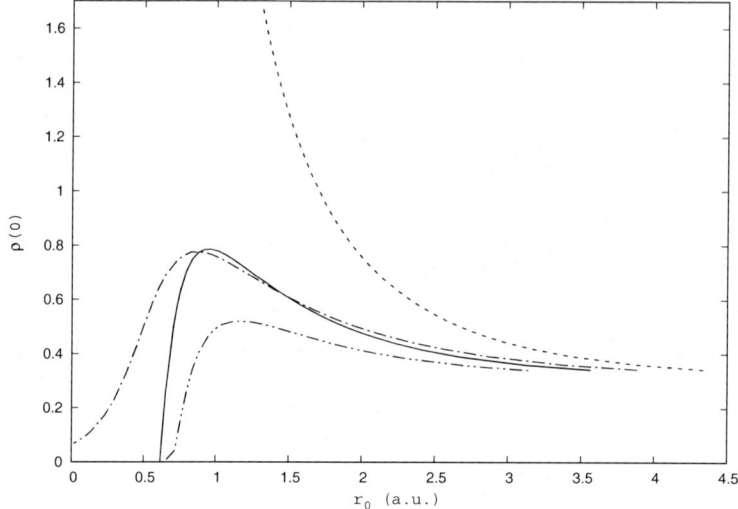

Figure 4. Electron density evaluated at the origin for the hydrogen atom confined by a impenetrable wall (dashed line), constant potential (solid line $U_0 = 0.5$ au), dielectric continuum (dot-dot-dashed line $\epsilon = 80$) and a parabolic potential (dot-dashed line $\omega = 0.5$ au).

tinuum potentials. In fact, when the confinement radius is close to the critical radius where the electron is detached by the confinement, $\rho(0)$ goes to zero. If we connect this result with that discussed for Q_0, it is clear that when $Q_0 \to 100$ then $\rho(0) \to 0$, it means that the electron is delocalized inside of the classical forbidden region. As a final comment, it is worth noting that the electron density evaluated at the origin has a behavior very similar to that exhibited by the kinetic energy (compare Figures 2 and 4).

3.2. Excited States of the Confined Hydrogen Atom

We have discussed results concerning to the ground state. However, exact analytical solutions presented above allow the calculation of properties for excited states of the hydrogen atom confined by impenetrable walls, a parabolic potential with $\omega = 0.5$ au, a constant potential with $U_0 = 0.5$ au and a dielectric continuum with $\epsilon = 80.0$.

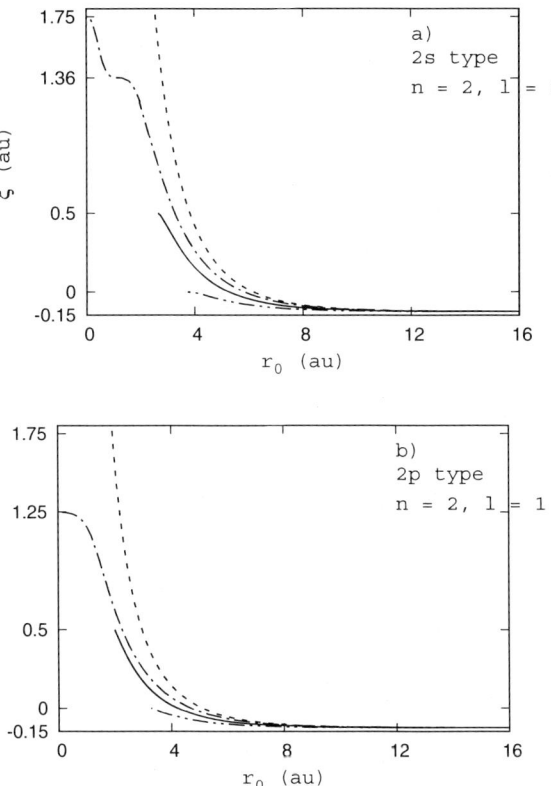

Figure 5. Total energy (ξ) as a function of r_0 for the states **a)** **2s** and **b)** **2p** of the hydrogen atom confined by impenetrable walls (dashed line), constant potential (solid line $U_0 = 0.5$ au), dielectric continuum (dot-dot-dashed line $\epsilon = 80$) and a parabolic potential (dot-dashed line $\omega = 0.5$ au).

The total energy for the states $2s$ and $2p$ as a function of the confinement radius is presented in Figure 5, where the response of these states to confinement potentials is contained in each plot. We must compare this figure with Figure 1 to appreciate differences between ground and excited states. From these figures it is clear that the effect of impenetrable walls on the total energy for all states is the same; this potential increases rapidly the total energy of the system. A constant potential and a dielectric continuum over excited states induce critical

Table 3. Critical radius, r_{crit}, where the confined hydrogen atom is ionized by two different confinements: continuum dielectric characterized by ϵ and a constant potential with $U_0=0$ au

State	Critical radius in atomic units	
	$\epsilon = 80.0$	$U_0 = 0.0$ au
$1s$	0.64428	0.72341
$2s$	3.60203	3.81161
$2p$	3.27726	3.29684
$3s$	9.10858	9.36753
$3p$	8.80863	8.85636
$3d$	7.18024	7.19792
$4s$	17.1469	17.3924
$4p$	16.8033	16.8779
$4d$	15.2739	15.3037

radii (r_{crit}) larger than the corresponding ground state. For some excited states, the critical radius is reported in Table 3 for the two potentials where an ionization can be observed for small confinement radii. In this table, the detachment energy of the electron is the same for $\epsilon = 80.0$ and $U_0 = 0.0$ au, where it is clear the impact of the shape of a constant potential since this potential suddenly reach its highest value at r_0, which is contrary to the behavior observed for the dielectric continuum and for that reason the r_{crit} is larger for the constant potential. Furthermore, we have found an interesting result related to the response of each state to the confinement. From this table r_{crit} exhibits the following ordering: $r_{crit}^{1s} < r_{crit}^{2p} < r_{crit}^{2s} < r_{crit}^{3d} < r_{crit}^{3p} < r_{crit}^{3s} < r_{crit}^{4d} < r_{crit}^{4p} < r_{crit}^{4s}$. This behavior is a consequence of the extension exhibited by each state; the more extended an orbital, the larger r_{crit}.

The parabolic potential deserves special attention since the energy associated to this confinement is drastically different to the other three potentials. One difference is found in the limit $r_0 \to 0$ where the energy goes to the value exhibited by the isotropic harmonic oscillator. Therefore, the hydrogen atom is not ionized when we apply the parabolic potential, and the energy does not rise indefinitely. The second interesting result obtained for this confinement is evidenced for the orbital $2s$ since there is a plateau before to reach the corresponding value of the harmonic oscillator when $r_0 \to 0$. This behavior is interesting and for that reason we computed all states from $n = 1$ to $n = 4$. All these energies as a function of r_0 are presented in Figure 6. From this figure, it is clear that we will obtain as many plateaus as the number of nodes in the wave function. Oscillations observed for kinetic and potential energies as a function of r_0 are the main reason of these plateaus. Each plateau is localized at the same position of a minimum of KE and a maximum of V. Such maxima and minima are a consequence of the behavior of the electron density when the hydrogen atom is under confinement. The third result for the parabolic confinement we want to mention is the degeneracy obtained for the limits $r_0 \to \infty$ and $r_0 \to 0$. For the first limit, the degeneracy is that observed for the hydrogen atom, and the second limit corresponds to the degeneracy of the isotropic harmonic oscillator.

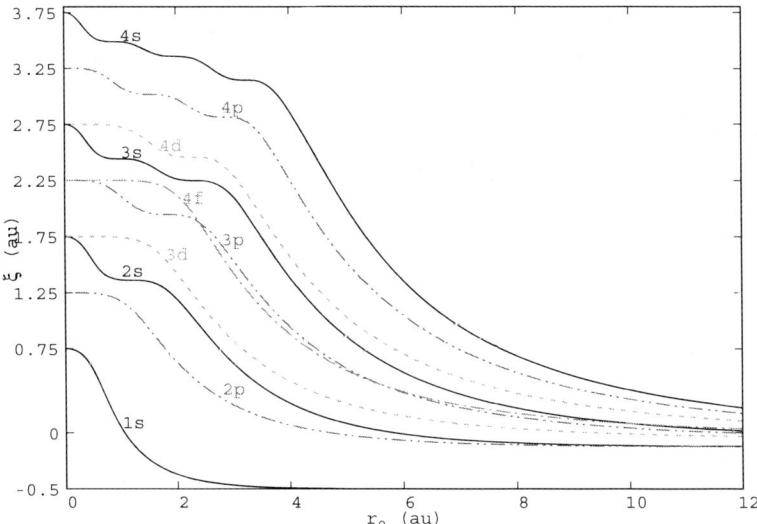

Figure 6. Total energy (ξ) as a function of r_0 for "s" type states (solid line), "p" type states (dot-dot-dashed line), "d" type states (dashed line) and "f" type state (dot-dashed line) of the hydrogen atom confined by a parabolic potential $\omega = 0.5$ au.

4. RITZ METHOD

In this section we do compare results from Ritz method with those obtained by the analytical solution. The Ritz method is based on the variational principle,

$$E[\psi] \leq \frac{\langle \psi | \hat{H} | \psi \rangle}{\langle \psi | \psi \rangle}, \tag{63}$$

with $E_0 = E[\psi_0]$, and ψ_0 corresponds to the wave function of the ground state.

In the Ritz procedure, ψ is proposed as a linear combination of k known spatial basis functions

$$\psi = \sum_{\nu=1}^{k} c_\nu \Phi_\nu(\mathbf{r}). \tag{64}$$

If this proposal to the wave function is inserted in equation (63) then the coefficients $\{c_\mu\}$ that minimize to the total energy must satisfy the algebraic equation

$$\mathbb{H}\mathbb{C} = \mathbb{S}\mathbb{C}\mathbb{E}, \tag{65}$$

with

$$S_{\mu\nu} = \int d\mathbf{r}\, \Phi_\mu^*(\mathbf{r}) \Phi_\nu(\mathbf{r}), \tag{66}$$

and

$$H_{\mu\nu} = \int d\mathbf{r}\, \Phi_\mu^*(\mathbf{r}) \hat{H} \Phi_\nu(\mathbf{r}) = \int d\mathbf{r}\, \Phi_\mu^*(\mathbf{r}) \left(-\frac{1}{2}\nabla^2 + v(r) \right) \Phi_\nu(\mathbf{r}). \tag{67}$$

Since the confinement potential exhibits spherical symmetry, each basis function is written as

$$\Phi_\nu(\mathbf{r}) = Y_{\ell_\nu, m_\nu}(\theta, \phi) R_\nu(r). \tag{68}$$

In this chapter we want to study the performance of GTOs, which are used in many quantum chemistry codes, over the hydrogen atom confined by four potentials. Thus, we have

$$R_\nu(r) = N_\nu r^{n_\nu - 1} e^{-\alpha_\nu r^2} \tag{69}$$

where α_ν is the Gaussian orbital exponent, n_ν is an integer and N_ν is a normalization constant given by the next relation

$$N_\nu = \left[\frac{2^{2n_\nu + 1}(2\alpha_\nu)^{n_\nu + \frac{1}{2}} n_\nu!}{\pi^{\frac{1}{2}}(2n_\nu)!} \right]^{\frac{1}{2}}. \tag{70}$$

The expression for the matrix elements of the kinetic energy is

$$T_{\mu\nu} = -\frac{1}{2}\delta_{\ell_\mu,\ell_\nu}\delta_{m_\mu,m_\nu} N_\mu N_\nu \left[\frac{n_\nu(n_\nu-1) - \ell_\nu(\ell_\nu+1)}{2(\alpha_\mu+\alpha_\nu)^{(n_\mu+n_\nu-1)/2}} - \frac{\alpha_\nu(1+2n_\nu)(n_\mu+n_\nu-1)}{2(\alpha_\mu+\alpha_\nu)^{(n_\mu+n_\nu+1)/2}}\right.$$

$$\left. + \frac{\alpha_\nu^2(n_\mu+n_\nu+1)(n_\mu+n_\nu-1)}{2(\alpha_\nu+\alpha_\mu)^{(n_\mu+n_\nu+3)/2}}\right] \Gamma\left(\frac{n_\mu+n_\nu-1}{2}\right). \tag{71}$$

And for the potential energy

$$V_{\mu\nu} = \delta_{\ell_\mu,\ell_\nu}\delta_{m_\mu,m_\nu} \left[\underbrace{\int_0^{r_0} r^2 R_\mu(r)\left(-\frac{Z}{r}\right)R_\nu(r)dr}_{\text{inside the cavity}} + \underbrace{\int_{r_0}^{\infty} r^2 R_\mu(r) V_c R_\nu(r)dr}_{\text{outside the cavity}}\right]$$

$$= \delta_{\ell_\mu,\ell_\nu}\delta_{m_\mu,m_\nu}\left(V_{\mu\nu}^{(\text{in})} + V_{\mu\nu}^{(\text{out})}\right), \tag{72}$$

with

$$V_{\mu\nu}^{(\text{in})} = -\frac{Z N_\mu N_\nu}{2(\alpha_\mu+\alpha_\nu)^{(n_\mu+n_\nu)/2}}\gamma\left(\frac{n_\mu+n_\nu}{2}, (\alpha_\mu+\alpha_\nu)r_0^2\right). \tag{73}$$

The expression for $V_{\mu\nu}^{(\text{out})}$ is:

- Parabolic potential

$$V_{\mu\nu}^{(\text{prb})} = \frac{\omega^2 N_\mu N_\nu}{4(\alpha_\mu+\alpha_\nu)^{(n_\mu+n_\nu+3)/2}}\Gamma\left(\frac{n_\mu+n_\nu+3}{2}, (\alpha_\nu+\alpha_\mu)r_0^2\right). \tag{74}$$

- Finite walls

$$V_{\mu\nu}^{(\text{fnt})} = \frac{U_0 N_\mu N_\nu}{2(\alpha_\mu+\alpha_\nu)^{(n_\mu+n_\nu+1)/2}}\Gamma\left(\frac{n_\mu+n_\nu+1}{2}, (\alpha_\nu+\alpha_\mu)r_0^2\right). \tag{75}$$

- Dielectric continuum

$$V_{\mu\nu}^{(\text{dlc})} = -\frac{Z N_\mu N_\nu}{2\epsilon(\alpha_\mu+\alpha_\nu)^{(n_\mu+n_\nu)/2}}\Gamma\left(\frac{n_\mu+n_\nu}{2}, (\alpha_\nu+\alpha_\mu)r_0^2\right). \tag{76}$$

For the matrix elements of the overlap

$$S_{\mu\nu} = \delta_{\ell_\mu,\ell_\nu}\delta_{m_\mu,m_\nu}\frac{N_\mu N_\nu}{2(\alpha_\mu+\alpha_\nu)^{(n_\mu+n_\nu+1)/2}}\Gamma\left(\frac{n_\mu+n_\nu+1}{2}\right). \tag{77}$$

The Gamma (Γ) function [27]

$$\Gamma(s) = \int_0^\infty t^{s-1}e^{-t}dt = \int_0^x t^{s-1}e^{-t}dt + \int_x^\infty t^{s-1}e^{-t}dt = \gamma(s,x)+\Gamma(s,x), \tag{78}$$

is involved in many expressions with

$$\gamma\left(\frac{1}{2},x\right) = \int_0^x t^{-\frac{1}{2}}e^{-t}dt = 2\int_0^{\sqrt{x}} e^{-y^2}dy = \sqrt{\pi}\operatorname{erf}(\sqrt{x}), \tag{79}$$

$$\gamma(1,x) = \int_0^x e^{-t}dt = 1 - e^{-x}. \tag{80}$$

In this case $\operatorname{erf}(z)$ is the error function [27]. Matrix elements with $s = \frac{3}{2}, 2, \frac{5}{2}, 3, \ldots$ can be evaluated using the next recurrence relation

$$\gamma(s,x) = (s-1)\gamma(s-1,x) - x^{s-1}e^{-x}. \tag{81}$$

It is important to mention that all integrals can be reduced to equations (79) and (80). With these definitions, the calculation of the upper incomplete Gamma function ($\Gamma(s,x)$) was performed through the equation (78). According to all the definitions exposed above, the argument of the Gamma function $s \to \frac{m}{2}$, with $m = 1, 2, 3, \ldots$, thus

$$\underbrace{\left(\frac{m}{2}-1\right)!}_{m \text{ even}} = \Gamma\left(\frac{m}{2}\right) = \underbrace{\frac{(m-1)!\sqrt{\pi}}{2^{m-1}\left(\frac{m-1}{2}\right)!}}_{m \text{ odd}}. \tag{82}$$

For impenetrable walls the procedure was practically the same, the main difference is the inclusion of a cutoff function on the radial part

$$R_\nu^{\text{imp}}(r) = N_\nu^{\text{imp}} r^{n_\nu-1}e^{-\alpha_\nu r^2}\left(1 - \frac{r}{r_0}\right). \tag{83}$$

4.1. Gaussian Functions in the Ritz Method versus Exact Results

To check the implementation of the matrix elements presented previously in the MEXICA-C code [12], the total energy of the ground state of the free hydrogen atom E as a function of the number (k) of GTOs is reported in Table 4. All GTOs used in the expansion were of type $1s$. The second column of this table corresponds to results reported by Huzinaga [30], and the third column corresponds to the results of our implementation. We see that our prediction is in

Table 4. Ground state energy for the *free* hydrogen atom as a function of number, k, of GTOs used in the Ritz method. All energies are in atomic units

k	Huzinaga [30]	This work
2	-0.485813	-0.48581272
3	-0.496979	-0.49697925
4	-0.499277	-0.49927840
5	-0.499809	-0.49980981
6	-0.499940	-0.49994555
7	-0.499976	-0.49998327
8	-0.499991	-0.49999453
9	-0.499997	-0.49999810
10	-0.499999	-0.49999929
Exact: -0.50000000		

good agreement with regard to results reported in the literature. Naturally, these results give us confidence of the implementation done in the MEXICA-C code.

Table 5 shows the energy of the $1s$ state of the hydrogen atom confined by a dielectric continuum, penetrable walls, a parabolic potential and impenetrable walls. For the four confinements r_0=1.0 and 4.0 au. From these results, there is an impressive result; GTOs plus cutoff function describe properly the $1s$ orbital when the hydrogen atom is confined by impenetrable walls since three GTOs are required to reach the exact value. The inclusion of the cutoff function plays a crucial role in the performance of this approach. For the rest of the potentials considered in this chapter, the GTOs show a poor performance, in particular, for the parabolic potential. In principle, GTOs have the correct asymptotic behavior for this potential, although it is evident that for this potential the Ritz

Table 5. Total energy of the hydrogen atom confined at $r_0=1.0$ au by a dielectric continuum ($\epsilon = 80.0$), penetrable walls ($U_0 = 0.0$ au), parabolic potential ($\omega = 1.0$ au) and impenetrable walls. All energies are in atomic units

	Energy			
k	$\epsilon = 80$	$U_0 = 0.0$ au	$\omega = 1.0$ au	Impenetrable
	$r_0=1.0$ au			
2	-0.111703	-0.107368	0.298441	2.373992
3	-0.122230	-0.117530	0.268678	2.373991
4	-0.128001	-0.123988	0.266892	2.373991
5	-0.128440	-0.124464	0.266653	2.373991
6	-0.128632	-0.124736	0.266568	2.373991
7	-0.129029	-0.124765	0.266462	2.373991
8	-0.129041	-0.124875	0.266239	2.373991
9	-0.129092	-0.124885	0.266338	2.373991
10	-0.129093	-0.124918	0.266343	2.373991
Exact	-0.129274	-0.125000	0.265165	2.373991
	$r_0=4.0$ au			
2	-0.485813	-0.485124	-0.476466	-0.476126
3	-0.495386	-0.495370	-0.482939	-0.481826
4	-0.497173	-0.497152	-0.483664	-0.483007
5	-0.497518	-0.497496	-0.485653	-0.483215
6	-0.497631	-0.497611	-0.485749	-0.483230
7	-0.497678	-0.497648	-0.485903	-0.483230
8	-0.497687	-0.497664	-0.486366	-0.483167
9	-0.497689	-0.497666	-0.486595	-0.483243
10	-0.497690	-0.497667	-0.486960	-0.483258
Exact	-0.497697	-0.497675	-0.488358	-0.483265

method needs of many basis set functions. The performance of GTOs to estimate the energy for dielectric continuum and constant potential is reasonably good, although with 10 GTOs is not possible to reach the exact value. From the same table it is clear that the role of the cutoff function for the impenetrable potential is less important for $r_0 = 4$ au since for this confinement 3 GTOs are not enough to obtain a good prediction. Curiously, the performance of the Ritz method with 10 GTOs is similar for all potentials, except for the parabolic potential. Thus, even when this basis set contains the asymptotic behavior demanded by the analytical solution its performance is not good when the Ritz method is used.

CONCLUSION

In this chapter we have systematized an approach to solve the Schrödinger equation for the hydrogen atom confined by four different potentials. For the inner region, where the Coulomb potential is present, the solution is the same for the four cases. For the region where the confinement potential is present the four cases exhibit different solution. However, all of them can be obtained from the Kummer's equation. By solving the Kummer's equation for each case we found the asymptotic behavior of the wave function of the confined hydrogen atom. Naturally, electronic properties of this atom exhibit different behavior which depends on the confined imposed. Whereas continuum dielectric and constant potential induce confinement radii where this atom is ionized, parabolic and impenetrable potentials do not induce such detachment, although the parabolic potential induce interesting results for total energies and its components since there are two limits in this potential; free hydrogen atom and isotropic harmonic oscillator. Gaussian type functions appear in a natural way in the solution of the parabolic potential coupled with the Coulomb potential. However, when a basis set based on GTOs is used with the Ritz method, the performance of this basis set is really poor to estimate the total energy for the hydrogen atom confined by this potential. This result was unexpected for us and it opens a discussion for results already published for confinements imposed by the harmonic oscillator, and variations of it, over atoms and molecules analyzed with GTOs [31, 32, 33].

ACKNOWLEDGMENTS

The authors thank the facilities provided by the Laboratorio de Supercómputo y Visualización en Paralelo at the Universidad Autónoma Metropolitana-Iztapalapa. Partial funding was provided by CONACYT, México, through the project FC-2016/2412. M. A. Martínez-Sánchez thanks CONACYT for the scholarship 574390.

REFERENCES

[1] Schrödinger E., *Collected papers on wave mechanics*. Chelsea Publishing Company, New York, 1982.

[2] Slater J. C., Atomic shielding constants. *Phys. Rev.*, 36:57–64, 1930.

[3] Boys S. F. and Egerton A. C., Electronic wave functions - i. a general method of calculation for the stationary states of any molecular system. *Proc. R. Soc. Lond. A.*, 200(1063):542–554, 1950.

[4] Ufimtsev I. S. and Martínez T. J., Quantum chemistry on graphical processing units. 1. strategies for two-electron integral evaluation. *J. Chem. Theory Comput.*, 4:222–231, 2008.

[5] Kussmann J. and Ochsenfeld C., Employing opencl to accelerate ab initio calculations on graphics processing units. *J. Chem. Theory Comput.*, 13:2712–2716, 2017.

[6] Miao Yipu and Merz Kenneth M., Acceleration of electron repulsion integral evaluation on graphics processing units via use of recurrence relations. *J. Chem. Theory Comput.*, 9:965–976, 2013.

[7] Szabo A. and Ostlund N. S., *Modern Quantum Chemistry: Introduction to Advanced Electronic Structure Theory*. Dover, New York, 1996.

[8] Parr R. G. and Yang W., *Density-Functional Theory of Atoms and Molecules*. Oxford University Press, Oxford, 1994.

[9] Michels A., De Boer J. and Bijl A., Remarks concerning molecural interaction and their influence on the polarisability. *Physica*, 4:981–994, 1937.

[10] Ludeña E. V., SCF calculations for hydrogen in a spherical box. *J. Chem. Phys.*, 66:468–470, 1977.

[11] Ludeña E. V., SCF Hartree-Fock calculations of ground state wavefunctions of compressed atoms. *J. Chem. Phys.*, 69:1770–1775, 1978.

[12] Garza J., Hernández-Pérez J.-M., Ramírez J.-Z., and Vargas R., Basis set effects on the Hartree-Fock description of confined many-electron atoms. *J. Phys. B: At. Mol. Opt. Phys.*, 45:015002, 2012.

[13] Sarsa A., Buendía E., and Gálvez F. J., Multi-configurational explicitly correlated wave functions for the study of confined many electron atoms. *J. Phys. B-At. Mol. Opt. Phys.*, 49:145003, 2016.

[14] Young T. D., Vargas R., and Garza J., A Hartree-Fock study of the confined helium atom: Local and global basis set approaches. *Phys. Lett. A*, 380:712–717, 2016.

[15] Gálvez F. J., Buendía E., and Sarsa A., Confinement effects on the electronic structure of m-shell atoms: A study with explicitly correlated wave functions. *Int. J. Quantum Chem.*, 117:e25421, 2017.

[16] Martínez-Sánchez M.-A., Rodriguez-Bautista M., Vargas R., and Garza J., Solution of the Kohn-Sham equations for many-electron atoms confined by penetrable walls. *Theor. Chem. Acc.*, 135(8):207, 2016.

[17] Duarte-Alcaráz F. A., Martínez-Sánchez M. A., Rivera-Almazo M., Vargas R., R. A. Rosas-Burgos, and Garza J.. Testing one-parameter hybrid exchange functionals in confined atomic systems. *J. Phys. B: At. Mol. Opt. Phys.*, 52:135002, 2019.

[18] Gorecki J. and Byers-Brown W., Padded-box model for the effect of pressure on helium. *J. Phys. B: At. Mol. Opt. Phys.*, 21:403–410, 1988.

[19] Ley-Koo E. and Rubinstein S., The hydrogen atom within spherical boxes with penetrable walls. *J. Chem. Phys.*, 71:351–357, 1979.

[20] Rodriguez-Bautista M., Díaz-García C., Navarrete-López A. M., Vargas R., and Garza J., Roothaan's approach to solve the hartree-fock equations for atoms confined by soft walls: Basis set with correct asymptotic behavior. *J. Chem. Phys.*, 143:34103, 2015.

[21] Rodriguez-Bautista M., Vargas R., Aquino N., and Garza J., Electron-density delocalization in many-electron atoms confined by penetrable walls: A Hartree-Fock study. *Int. J. Quantum Chem.*, 118:e25571, 2018.

[22] Tomasi J., Mennucci B., and Cammi R., Quantum mechanical continuum solvation models. *Chem. Rev.*, 105:2999–3093, 2005.

[23] Cramer C. J. and Truhlar D. G., Implicit solvation models: Equilibria, structure, spectra, and dynamics. *Chem. Rev.*, 99:2161–2200, 1999.

[24] Jortner J. and Coulson C. A., Environmental effects on atomic energy levels. *Mol. Phys.*, 24:451–464, 1961.

[25] Martínez-Sánchez M.-A., Aquino N., Vargas R., and Garza J., Exact solution for the hydrogen atom confined by a dielectric continuum and the correct basis set to study many-electron atoms under similar confinements. *Chem. Phys. Lett.*, 690:14–19, 2017.

[26] Aquino N., Campoy G. and Montgomery H. E., Highly accurate solutions for the confined hydrogen atom. *Int. J. Quantum Chem.*, 107:1548–1558, 2007.

[27] Abramowitz M. and Stegun I. A., *Handbook of Mathematical Functions With Formulas, Graphs, and Mathematical Tables*. National Bureau of Standards, United States of America, Washington, D.C., 1964.

[28] Henning C., Baumgartner H., Piel A., Ludwig P., Golubnichiy V., Bonitz M., and Block D., Ground state of a confined yukawa plasma. *Phys. Rev. E*, 74:056403, 2006.

[29] Wolfram Research, Inc. Mathematica, Version 11.2.0.0. Champaign, IL, 2019.

[30] Huzinaga S., Gaussian-type functions for polyatomic systems. I. *J. Chem. Phys.*, 42(4):1293–1302, 1965.

[31] Sako T. and Diercksen G. H. F., Confined quantum systems: dipole polarizability of the two-electron quantum dot, the hydrogen negative ion and the helium atom. *J. Phys. B-At. Mol. Opt. Phys.*, 36:3743–3759, 2003.

[32] Sako T., Yamamoto S. and Diercksen G. H. F., Confined quantum systems: dipole transition moment of two- and three-electron quantum dots, and of helium and lithium atoms in a harmonic oscillator potential. *J. Phys. B-At. Mol. Opt. Phys.*, 37:1673–1688, 2004.

[33] Choluj M. and Bartkowiak W., Electric properties of molecules confined by the spherical harmonic potential. *Int. J. Quantum Chem.*, 119:e25997, 2019.

INDEX

A

amplitude, viii, 53, 57, 72, 73, 74, 79, 83
asymptotics, 90
atoms, viii, 101, 102, 103, 104, 130, 131, 132

B

boson, 79, 80, 82, 83, 86
bounds, 114

C

calculus, 2, 51
causality, 54
CERN, 95
classes, 15
conception, 10, 94
configuration, 104
confinement, ix, 97, 101, 103, 104, 105, 107, 111, 113, 114, 115, 116, 118, 119, 121, 122, 123, 124, 127, 129
conservation, 81
construction, 57, 65, 71, 73, 87, 115
correlation, 56, 57
correlation function, 56, 57
Coulomb interaction, 110
covering, 1, 2
critical value, viii, 54, 87, 92, 93

D

DEL, 1
derivatives, 56, 63
detachment, 121
differential equations, vii, 1, 2
Dirac equation, 93
distribution, 2

E

electron, viii, ix, 101, 102, 103, 104, 108, 111, 118, 119, 120, 121, 123, 130, 131
electronic structure, ix, 102
energy, 102, 106, 108, 109, 111, 113, 114, 115, 116, 117, 118, 120, 121, 122, 123, 125, 127, 128, 131
environment(s), viii, 101, 103

F

field theory, 53, 54, 57, 95, 97, 98
formula, 64

G

gravity, 55, 96

H

Hartree-Fock, 104, 131
helium, 131, 132
hydrogen, vii, ix, 101, 102, 103, 105, 108, 110, 113, 114, 116, 118, 120, 122, 123, 124, 127, 129, 131, 132

I

integration, 10, 13, 16, 58, 76, 80, 85
ionization, 104
iteration, 74, 76, 84

K

Kohn-Sham equations, 131

L

laws, 2
lithium, 132
localization, 119

M

mapping, 19
mass, 72, 74, 75, 82, 86, 87
matrix, 65, 86, 125
matter, 96
models, vii, viii, 54, 55, 56, 57, 71, 76, 94, 131
molecules, 102, 104
momentum, viii, 54, 61, 62, 69, 74, 82, 83, 87, 92, 94

N

nodes, 111
non-linear equations, 73
normalization constant, 102, 106, 107, 111, 112
NRC, 53
nuclear charge, 93
nucleus, 99, 103

O

ordinary differential equations, 1, 2
OSC, 1
oscillation, vii, 1, 2, 14
overlap, 125

P

partial differential equations, vii, viii, 2
particle physics, 96
partition, 10, 11
pathology, 53
physics, 56
poor performance, 127
principles, 54
project, 129
propagators, 57, 58, 90

Q

quantum chemistry, 124
quantum electrodynamics (QED), 55, 95, 96
quantum field theory, vii, viii, 54, 55, 56, 61, 98
quantum theory, 93
quarks, 56

R

radius, ix, 101, 103, 104, 111, 113, 114, 115, 116, 119, 122
real numbers, 3, 12, 105
recurrence, 106, 130
renormalization, 54, 56, 72, 73, 75, 79, 82, 83, 87
repulsion, 130
researchers, vii, 1, 2
residue, 54
response, 93
restoration, 65
restrictions, ix, 102
roots, 78

S

scalar field, viii, 53, 54, 56, 57, 58, 60, 71, 76, 79, 97, 98
scalar field theory, 97
scaling, 110
signs, 86
simulation, 61
solution, vii, viii, ix, 7, 13, 15, 16, 17, 23, 24, 25, 34, 43, 45, 46, 47, 48, 49, 50, 53, 54, 57, 58, 63, 64, 67, 69, 73, 74, 76, 77, 78, 82, 83, 87, 89, 90, 91, 101, 102, 104, 105, 106, 107, 108, 109, 111, 113, 114, 115, 116, 123, 128, 129, 132
solvation, 131
spin, 56
stability, 2, 15, 16

standard model, 95
structure, viii, 54, 55, 57, 64, 71, 92, 98, 131
symmetry, 64, 65, 106, 124

T

techniques, 103
total energy, 103, 106, 113, 114, 118, 121, 124, 126, 129
transformation, 107

V

vector, 65

Related Nova Publications

CHAOS FOR LINEAR OPERATORS AND ABSTRACT DIFFERENTIAL EQUATIONS

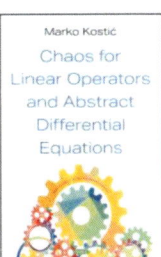

AUTHOR: Marko Kostić

SERIES: Mathematics Research Developments

BOOK DESCRIPTION: The theory of linear topological dynamics is a rapidly growing field of research over the last three decades or so. This book presents a survey of recent results of the author obtained in this field during the period 2016-2019.

HARDCOVER ISBN: 978-1-53616-895-2
RETAIL PRICE: $230

PATHS IN COMPLEX ANALYSIS

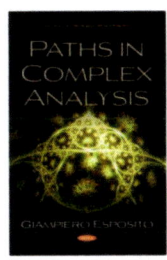

AUTHOR: Giampiero Esposito

SERIES: Mathematics Research Developments

BOOK DESCRIPTION: The book is unique both for the selection of topics and for the readable access that it offers to the otherwise too large landscape of modern complex analysis.

SOFTCOVER ISBN: 978-1-53617-057-3
RETAIL PRICE: $82

To see a complete list of Nova publications, please visit our website at www.novapublishers.com

Related Nova Publications

Mathematical Modeling of Real World Problems: Interdisciplinary Studies in Applied Mathematics

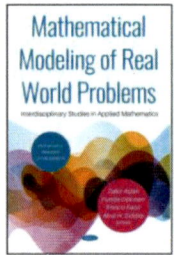

Editors: Zafer Aslan, Funda Dökmen, Enrico Feoli, and Abul H. Siddiqi

Series: Mathematics Research Developments

Book Description: Data mining provides avenues for proper understanding of real world problems. For researchers interested in data mining and new applications, this book is a multidisciplinary 'handbook' in data processes, engineering and medical applications.

Hardcover ISBN: 978-1-53616-267-7
Retail Price: $230

Understanding Banach Spaces

Editor: Daniel González Sánchez

Series: Mathematics Research Developments

Book Description: This book focuses on the study of several properties of Banach spaces applied to diverse problems in functional and numerical analysis. Many problems in science, engineering and other disciplines can be expressed in the form of equations, inequalities or systems of equations using mathematical modelling.

Hardcover ISBN: 978-1-53616-745-0
Retail Price: $270

To see a complete list of Nova publications, please visit our website at www.novapublishers.com